Make Science Fun
EXPERIMENTS

First published in 2017 by New Holland Publishers
Sydney

Level 1, 178 Fox Valley Road, Wahroonga, NSW 2076, Australia

newhollandpublishers.com

ISBN 9781742570716

Managing Director: Fiona Schultz
Publisher: Monique Butterworth
Project Editor: Liz Hardy
Designer: Andrew Davies
Book Reviewer: Samuel Capell
Production Director: James Mills-Hicks
Printed in China

10 9 8 7 6 5 4 3 2

Keep up with New Holland Publishers:

 NewHollandPublishers
 @newhollandpublishers

Thanks

A special thanks to everyone mentioned below for assisting with trialling the experiments, making suggestions and helping with photography: Thalia Angelopoulos, Haven Barr, Wade Barr, Harry Bath, Eliana Bootes, Arden Cannone, Neve Casey, Casey Corcoran, Riley de Courcey, Angus Cross, Kaylee Emery, Andre Gardiman, Charli Gibbens, Chloe Gleeson, Sienna Healey, Summer Healey, Janine and Todd Healey, Alana Hillman, Joanne and Craig Hillman, Kaia James, Talita James, Nigel and Elisa James, Zoe Kongonis, Tabitha Lay, Charlotte Lockett, Evie Pierce, Ella Sandeman, Ben Sheather, Nicholas Sparks, Colby Step, Arabella Stevens, Mikalah Strickling, Samuel Strickling, Jessica Swaine, Lily Thomson, Timothy Verster, Darcy Walker, Paula and Warren Walker, Cameron Wallace and Ishbel Wood.

Contents

Introduction

I love science and so do my kids!

Did you know that most science-experiment books don't actually include many experiments at all! They mostly contain a whole heap of fun and interesting science activities. These might be enjoyable to do and help kids learn science, but this doesn't make them experiments!

A science experiment is different from a science activity in that it sets out to answer a question or solve a problem using fair and controlled tests. You need to take measurements, make observations and control variables. Sounds hard, doesn't it! Don't worry, by the time you've completed all the experiments in this book you'll be ready to apply to do a PhD (or at least have a better understanding of the scientific method!).

This book is perfect for 8 to 15 year olds and contains a whole bunch of experiments with lots of sample results to help keep you on track! Most of the equipment and materials can be found around the home, however one piece of equipment that will help in many of the experiments is a thermometer. These can be purchased cheaply online.

Safety Note

By its inherent nature, science can be dangerous. Children should be supervised at all times when they are doing any of the activities described in this book. Children under the age of three shouldn't do any of the activities due to the numerous choking hazards and other dangers.

I would advise you to carefully read through the entire activity before attempting it. This will help make sense of what you are doing and make you more aware of any possible dangers.

Time required

An approximate time in hours to carry out the experimental procedure and complete the written report.

Difficulty

A difficulty level out of 5. The easiest experiments given a 1, the most challenging a 5.

Dangers

Some of the obvious dangers involved are stated. The responsible, supervising adult will need to make their own decisions regarding safety.

Materials and equipment

A basic list of the materials and equipment required.

Notes regarding science fair projects

In my role as a science teacher I have marked and graded hundreds of science fair projects over the past 21 years. In fact, my number one video on my YouTube Channel 'Make Science Fun' is 'How to get a grade A in your Science Fair Project'! So here are a few hints to do well:

1. Actually do some form of experiment. Don't just do a demonstration, survey or questionnaire. You literally need to set up some equipment and take measurements over time. You may need to have a control component in your experiment and you must have a clear independent variable and a dependent variable.
2. Choose a topic that has some form of real life application.
3. Choose a topic that is at least slightly interesting.
4. Do the experiment yourself and include some photographic or video evidence that you completed it. Make sure you include yourself in the photos!

There are lots of great examples in this book. Make sure you do the experiment yourself though, the sample results are just to guide you and keep you on track!

The best way to explain the features of a great project is to go through an actual example with explanations of the headings and some sample responses.

Note well: Your school system may require slightly different headings, but the essence of the project will be the same across all school systems and countries.

Sample science fair project

Title

A short statement to generate some interest and to give a hint what the project is about.

Example: Sugar content and the fermentation of yeast

Summary

A paragraph giving an overall picture of what the project is about and its applications.

Example: Yeast is super important. These little microbes are what make bread dough rise as they produce carbon dioxide by eating sugar. The environmental conditions such as warmth and the amount of sugar available can have a dramatic effect the fermentation rate. In this experiment we explore the effect of changing the concentration of sugar on how quickly the yeast ferments.

Question

Finding the relationship between an independent variable and a dependent variable.

An easy way to form a question is: How will changing _____ affect _____ ?

Example: How will changing the amount of sugar affect the fermentation activity of yeast?

Research

Find out a number of facts about your topic from the internet, books, journals or textbooks. Make sure you reference the sources from where you got the information!

Example: Yeast are unicellular microorganisms and are part of the fungus family.

Yeast convert sugar into ethanol and carbon dioxide by a process called fermentation.

Fermentation has been used for thousands of years to make bread rise.

The biological activity of yeast is determined by the temperature of its environment and the amount of sugar present.

Reference: www.britannica.com/science/yeast-fungus

Hypothesis

A testable statement (like a scientific guess!) based on your own past experience and research. It is okay if your hypothesis turns out to be wrong!

Example: I think the higher the concentration of the sugar the greater the rate of carbon dioxide produced by the yeast.

Variables

A variable is a part of the experiment which can be changed.

Independent variable – this is the variable that you choose to change throughout the experiment.

Example: The independent variable is the amount of sugar in the yeast solution.

Dependent variable – this is the aspect of the experiment that depends on or changes as a result of changing the independent variable.

Example: The dependent variable will be how much carbon dioxide is produced.

Controlled variables – all the variables/aspects of the experiment that must be kept the same.

Example: The controlled variables include:

- The starting amount of yeast
- The starting temperature
- The starting amount of water
- The length of time of fermentation

The Control

This idea is a bit tricky. The control is the test in an experiment that acts as a baseline. The control shows what happens when there are no changes. Not all experiments will have a specific control.

Example: The control will be yeast growing without any sugar added at all.

It can be helpful to plan the experiment with a grid approach:

Controlled variable	Controlled variable	Controlled variable
The initial amount of yeast (7 g sachet)	The amount of water (250 ml)	The initial temperature of the mixture (33°C)
Controlled variable	**Independent variable**	**Controlled variable**
The type of yeast (dried powder yeast)	The number of tablespoons of sugar in 250 ml of water (0, 1, 2, 3, 4, 5)	The type of bottle holding the mixture
Controlled variable	**Controlled variable**	**Controlled variable**
The ongoing temperature of the room	The type of balloon on the bottle (they all need to have the same elasticity)	Whatever other variable you think needs to be controlled...

Sample science fair project

In this diagram, the independent variable is placed in the centre of the grid. The variables to keep the same, or control, are placed around the centre. They specify the controlled variables and the quantities and situations that will be used.

Materials and Procedure

A list of materials and a numbered list giving the instructions for the experiment.

Example Materials

- 6 plastic water bottles
- 6 sachets of dried yeast
- 6 balloons
- 1 cup of sugar
- Water

Example Procedure

1. Put 250 millimetres of warm (33 degrees Celsius) tap water into each of 6 small plastic bottles.

2. Add a 7 gram sachet (or similar) of yeast to each bottle.

3. To the first bottle, attach a balloon by stretching the opening of it over the top of the bottle. This bottle without sugar will be the **control**.

4. Add 1 tablespoon of sugar to the next bottle and attach a balloon.

5. Add 2 tablespoons of sugar to the next bottle and so on until the sixth bottle has 5 tablespoons of sugar and a balloon over the top.

6. Give each bottle a good swirl to dissolve and mix the sugar, yeast and water together.

7. Label each bottle with the number of tablespoons of sugar it contains.

8. Take regular photos of the balloons inflating and measure the diameter of each balloon at the end of 30 minutes.

Results – Table/Graphs/Photographs

Include photos of you in action – for example adding the ingredients to the bottles and looking like you're having fun! Include photos of all the bottles and balloons at the end of the experiment.

Include all your results in a table. It is okay to have handwritten results in the table.

Sample results

Number of tablespoons of sugar	Diameter of balloon after 30 minutes (millimetres)
0	0
1	78
2	85
3	105
4	100
5	45

Graphs are usually line graphs (known as x-y scatter graphs in Excel) or column graphs. In this case I think a line graph is best.

Sample Graph

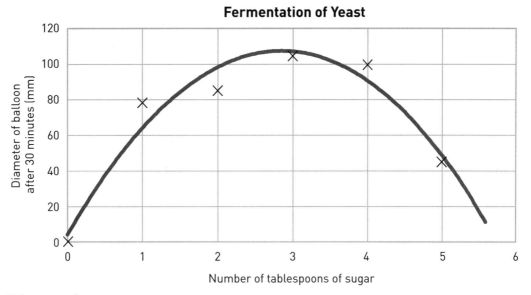

Discussion

Here is your chance to shine! Spend time thinking about the results. Can you draw any clear conclusions? Did one amount of sugar promote production of carbon dioxide gas more than the others? Were there any trends? What about unexpected results? Make some suggestions how the experiment could be improved. What are the implications of your results to real life applications?

Example: It was quite clear that the optimum amount of sugar was 3 tablespoons. With no sugar, the yeast did not produce any carbon dioxide at all. I was very surprised to see the gas production reduce so dramatically after 4 tablespoons of sugar. It's as if the solution became too sweet for the yeast to function. Some improvements would be to try 2½ and 3½ tablespoons of sugar and also to stand the bottles in a pan of warm water to help keep the temperature constant. Next time I make bread I'm going to add 3 tablespoons of sugar to 250 ml of water as a starter! It would be interesting to find out how temperature affects the biological activity of the yeast.

Conclusion

The conclusion itself will be a short statement answering the stated aim of the experiment.

Example: The ideal amount of sugar, which produced the most carbon dioxide, is about 3 tablespoons in 250 ml of water.

Use the following illustration as a guide to assist setting out your science fair project.

Science Project Display Board

For consistency, each experiment in this book has the features set out in this example.

The idea of this book is to give you *lots* of help at the beginning, and then progressively get you to plan and do more and more of the experiments yourself. So to start with, the early experiments have basically given you *all* the information you need, but as the book progresses, the amount of detail gets less and less, allowing you to do more and more yourself.

Now it's your turn to do some science! Have fun and stay safe.

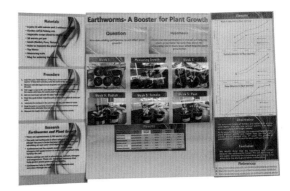

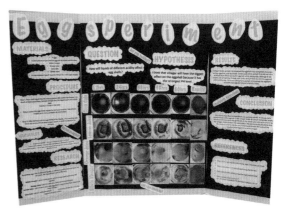

Sample science fair project

1. Runny honey

Does temperature affect how viscous honey is?

TIME REQUIRED – 2 hours

DIFFICULTY – 3 out of 5

DANGERS – Be very careful with hot water

Materials and Equipment

- Jar of honey
- 2 bowls
- Hot water
- Spoon
- Sticky tape
- Large funnel
- Spatula
- Thermometer
- Stopwatch
- Ice

Summary

What's the stickiest, thickest and most viscous liquid you can think of? Probably honey, right? Well, let me tell you about a *really* viscous liquid – bitumen (or pitch, as it's also known). This black, sticky stuff is used to make roads. In 1927, a scientist in Australia set up a funnel full of bitumen and displayed it in a glass box. It's *so* viscous the funnel only drips a drop of bitumen about once every eight years! The university where it's kept managed to capture a drip happening on camera – you can check it out on YouTube!

Some other viscous liquids include motor oil, vegetable oil and glucose syrup. Viscosity is a measure of a liquid's resistance to flow.

In this experiment we explore the viscosity of honey.

Question

How will changing the temperature of honey affect how runny (viscous) it is?

Research

For you to do!

What is viscosity? _____

What is honey? _____

What generally happens to the viscosity of liquids at higher temperatures? _____

_____

Explain why the viscosity changes with temperature _____

_____

Hypothesis

The higher/lower the temperature of the honey, the more viscous it is.

Circle your best guess

Procedure

1. Stand a jar of honey in a plastic bowl containing hot water for 15 minutes.

2. Mix the honey around inside the jar with a spoon until the honey has uniformly heated up. Measure and record the temperature of the honey in the results table.

3. Put some sticky tape on the bottom of a funnel, and pour the entire jar of honey into the funnel.

4. Remove the sticky tape and time how long it takes for the honey to empty out of the funnel and back into the jar. Record the time in the results table.

5. Put the jar of honey back into the bowl of hot water (it would have cooled down a bit) mix the honey around and then measure its temperature again.

6. Repeat steps 3–5 several times.

7. Eventually you can add some ice to the water in the bowl to reach even lower temperatures.

1. Runny honey

Variables

Controlled variable	Controlled variable	Controlled variable
The amount of honey (one jar full)	The size of the funnel	Holding the funnel still (not jiggling it up and down)
Controlled variable	**Independent variable**	**Controlled variable**
The type of honey	The temperature of the honey	The material the funnel is made from
Controlled variable	**Controlled variable**	**Controlled variable**
Any other variable you think needs to be controlled...	_____ _____	_____ _____

Dependent Variable – The time it takes for the honey to flow out of the funnel.

Sample Results

Temperature (°C)	Time (seconds)	Observations
52	12	Extremely runny, flows very fast
47	22	Very low viscosity
40	24	Runny and thick at the same time
38	29	I cooled the water down with some ice because the temperature of the honey was not changing
31	56	Not as runny
28	67	Honey is getting much thicker
25	87	Honey is sticking to the thermometer
21	144	Much higher viscosity than at the start
16	221	Very thick and sticky. Very viscous.

Sample Graph

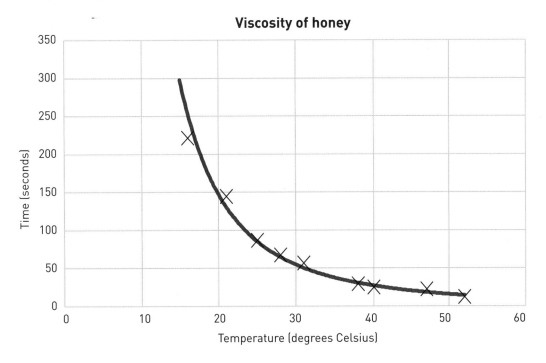

Sample Discussion

I was very surprised to see how quickly the honey flowed at 52°C. At the lower temperatures the honey was very sticky and flowed slowly. Clearly the viscosity of the honey is affected by the temperature. To get honey out of a jar quickly, warm it up!

Sample Conclusion

The lower the temperature of the honey the more viscous it is.

Your Results

Temperature (°C)	Time (seconds)	Observations

Your Graph

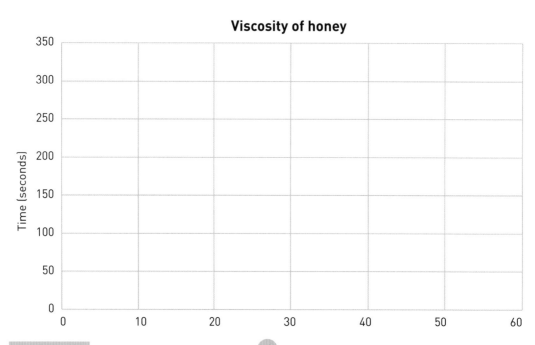

Viscosity of honey

1. Runny honey

Your Discussion

Identify the clear trends_____

_____

_____

Discuss any unexpected results_____

_____

_____

Suggest how the experiment could be improved_____

_____

_____

What are the implications of your results to real life applications?_____

_____

_____

Your Conclusion

_____

_____

Further Experiment

Try testing another viscous liquid such as vegetable oil.

1. Runny honey

2. Bouncy balls
Do hot balls bounce higher?

TIME REQUIRED – 2 hours
DIFFICULTY – 3 out of 5
DANGERS – Be very careful with hot water

Summary

A sport I enjoyed playing when I was young and fit was called squash. Two opponents entered a small arena and using squash racquets hit a small black rubber ball against the walls. Before starting, one of the players would always take the time to roll and squish the ball on the ground with their shoe. This warmed up the ball, which apparently made it more bouncy! Why not do a scientific experiment to find out if warmed up rubber balls are actually bouncier?

Materials and Equipment
- Ice
- Hot water
- Thermometer
- Rubber ball
- Tape measure

Question

How will changing the temperature of a rubber ball affect how bouncy it is?

Research

What is rubber? _____

For you to do!

_____

_____

_____

How does temperature affect the elasticity of rubber? _____

_____

_____

_____

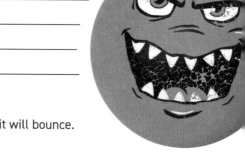

Hypothesis

The higher the temperature of a rubber ball the lower/higher it will bounce.

Circle your best guess

Procedure

1. Using the ice and water make up a bowl of water that has a temperature of 10 degrees Celsius.

2. Put a rubber ball in this water for 3 minutes.

3. Drop the ball from 1-metre high onto a hard surface and record how high it bounces. Record this value in the results table.

4. Repeat by dropping the ball again from 1-metre high and record how high it bounces. Record this value and then calculate the average bounce height.

5. Using hot water from the tap and some cold water, make a new bowl of water which has a temperature of 20 degrees Celsius. Drop the ball from 1-metre high and record the height it bounces. Repeat.

6. Using the hot water from the tap try at least three more temperatures such as 30, 40 and 50 degrees Celsius.

Variables

Controlled variable	Controlled variable	Controlled variable
The type of ball used	The size of the ball	The material the ball is made from
Controlled variable	**Independent variable**	**Controlled variable**
The surface the ball is dropped onto	The temperature of the ball	The initial height of the drop (1 metre)
Controlled variable	**Controlled variable**	**Controlled variable**
Any other variable you think needs to be controlled...	_____ _____	_____ _____

Dependent Variable – The height the ball bounces to.

Sample Results

Average

Tempera-ture (°C)	Height of bounce (cm)		
	Bounce 1	Bounce 2	Average
10	20	24	22
20	25	27	26
30	30	32	31
40	36	36	36
50	43	45	44

The Average = (Bounce 1 + Bounce 2) ÷ 2

Sample Graph

Bounce height of ball dropped from 50cm

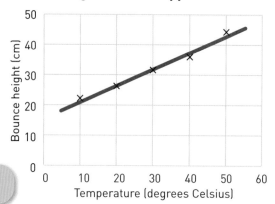

Sample Discussion

The results were more dramatic than I expected! Bounce heights were noticeably affected by the temperature. The hotter the ball got, the higher it bounced! At cold temperatures the rubber particles making up the ball are closer together and so are not as elastic.

Sample Conclusion

The bounce height is proportional to the temperature. As the temperature increases, the bounce height increase.

Now it's your turn!

Your Results

Temperature (°C)	Height of bounce (cm)		
	Bounce 1	Bounce 2	Average

2. Bouncy balls

Your Graph

Bounce height of ball dropped from 50cm

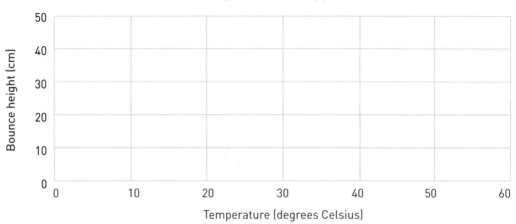

Your Discussion

Identify the clear trends_____

_____

Discuss any unexpected results_____

_____

Suggest how the experiment could be improved_____

_____

What are the implications of your results to real life applications?_____

_____

Your Conclusion

_____

_____

Further Experiments

Are golf balls affected by temperature?

Does the type of surface a ball bounces on affect rebound height?

2. Bouncy balls

3. Vortex bottle
The quickest way to empty a bottle of water

TIME REQUIRED – 2 hours
DIFFICULTY – 2 out of 5
DANGERS – No obvious dangers

Materials and Equipment
- Plastic bottle
- Water
- Stopwatch

Summary

Imagine it's a really hot day and you want to cool yourself off as quickly as possible. You grab a cold bottle of water and hold it upside down over your head. What do you think would be the quickest way to empty it? Shaking it up and down? Swirling it in a circle? Why not try an experiment to find out!

Question

What's the quickest way to empty a bottle of water?

Hypothesis

I think a bottle full of water will empty the fastest when it is shaken/held still/swirled/ _____

Procedure

1. Fill a 600 ml plastic bottle with water, turn it upside down and hold it steady. Time how long it takes for the water to pour out. Repeat a total of five times. This part of the experiment will be the control.

2. After refilling, empty the bottle by shaking it up and down vigorously. Time how long it takes and repeat this a total of five times.

3. Now try some other techniques to empty it such as shaking it side to side and swirling the bottle to create a vortex (swirly whirlpool!).

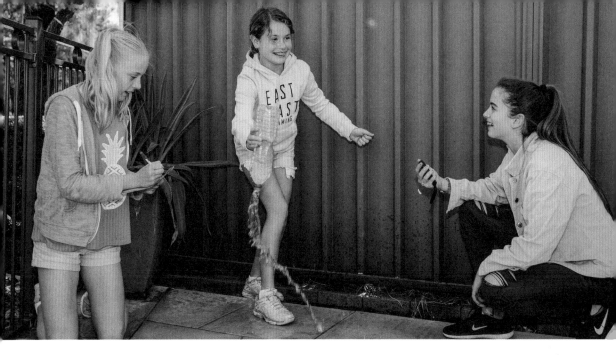

Variables

Controlled variable	Controlled variable	Controlled variable
The bottle used (600 ml plastic bottle)	How much water is used (Full – 600 ml)	The temperature of the water (cold tap temperature)
Controlled variable	**Independent variable**	**Controlled variable**
The same person to use the same emptying motion each time	The method of emptying the bottle (held steady, shaken up and down, shaken sideways, swirled)	Make sure to do it on planet Earth each time! We want to make sure we keep gravity the same!
Controlled variable	**Controlled variable**	**Controlled variable**
Any other variable you think needs to be controlled...	_____ _____	_____ _____

Dependent Variable – The time it takes for the bottle to empty.

The Control

Holding the bottle steady without shaking it.

To calculate the average – add the five times together and divide by five.

Sample Results

How the bottle is emptied	Time (seconds)					
	1	2	3	4	5	Average
Held steady	7.1	7.0	8.2	6.9	7.9	7.4
Shaken up and down	8.9	7.1	8.4	7.8	7.3	7.9
Shaken side to side	5.4	6.6	6.0	6.3	5.9	6.0
Swirled to get a vortex	5.7	6.5	5.6	5.3	4.7	5.6

Sample Graph

As there is no numerical relationship in the ways in which the bottled is emptied, a column graph is the best type of graph to use in this case.

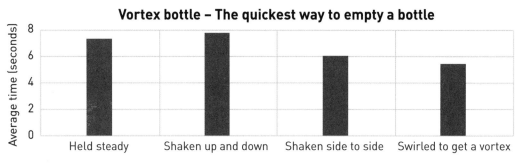

Vortex bottle – The quickest way to empty a bottle

Sample Discussion

Shaking the bottle took the longest to empty. Swirling the bottle causes it to empty the fastest. The vortex allows air into the bottle through the centre of the vortex, which then helps push out the water.

To improve this experiment you could try using a larger bottle. This would increase the time of emptying, which would mean reaction-time errors for starting and stopping would not be as significant.

Sample Conclusion

The fastest way to empty a bottle of water is to swirl it and produce a vortex.

Your Results

How the bottle is emptied	Time (seconds)					
	1	2	3	4	5	Average
Held steady	7.1	7.0	8.2	6.9	7.9	7.4
Shaken up and down	8.9	7.1	8.4	7.8	7.3	7.9
Shaken side to side	5.4	6.6	6.0	6.3	5.9	6.0
Swirled to get a vortex	5.7	6.5	5.6	5.3	4.7	5.6
Your own technique 1 _____ _____						
Your own technique 2 _____ _____						

Your Graph

Vortex bottle – The quickest way to empty a bottle

Average time (seconds)

Held steady   Shaken up and down   Shaken side to side   Swirled to get a vortex

Method to empty bottle

Your Discussion

Identify the clear trends_____

_____

_____

_____

Discuss any unexpected results_____

_____

_____

_____

Suggest how the experiment could be improved_____

_____

_____

_____

What are the implications of your results to real life applications?_____

_____

_____

_____

_____

Your Conclusion

_____

_____

Further Experiments

Does the shape of the bottle affect how quickly it empties?
Does the type of liquid affect how quickly the bottle empties?

4. Flip my bottle
The perfect amount of water for a successful bottle flip

TIME REQUIRED - 3 hours
DIFFICULTY - 2 out of 5
DANGERS - No obvious dangers

Materials and Equipment
- A plastic bottle suitable for bottle flipping
- Measuring jug
- Water

Summary

2016 was the year of the bottle flip! Young people loved the challenge of flipping a plastic bottle containing a small amount of water so that the bottle spins once in the air and lands firmly on a table. Practice was the key to success (as well as a bit of luck!). An interesting question is what is the best amount of water in the bottle for a successful bottle flip?

Question

What is the optimum amount of water for a successful bottle flip?

Hypothesis

The optimum percentage of water in a bottle for a successful flip is _____%.

Procedure

1. Find a plastic bottle generally used for bottle flipping.
2. Without adding any water, flip it 50 times onto a surface and record the number of successful flips.
3. Add 50 ml of water to the bottle and flip it 50 times onto a surface. Record the number of successful flips.
4. Repeat step 3 until the bottle is full! (That's a lot of flips!)

Variables

Controlled variable	Controlled variable	Controlled variable
The actual motion of the flip	The bottle used (500 ml bottle)	The surface the bottle is flipped onto
Controlled variable	**Independent variable**	**Controlled variable**
The person doing the flipping	The amount of water in the bottle 50 ml, 100 ml,150 ml...and so on	The type of liquid in the bottle (water)
Controlled variable	**Controlled variable**	**Controlled variable**
_____ _____	_____ _____	_____ _____

Dependent Variable – A successful bottle flip!

Sample Results

Amount of water (ml)	Percentage of water in bottle (%)	Total number of successful flips
0	0	0
50	10	0
100	20	1
150	30	10
200	40	30
250	50	20
300	60	30
350	70	24
400	80	25
450	90	16
500	100	14

Sample Graph

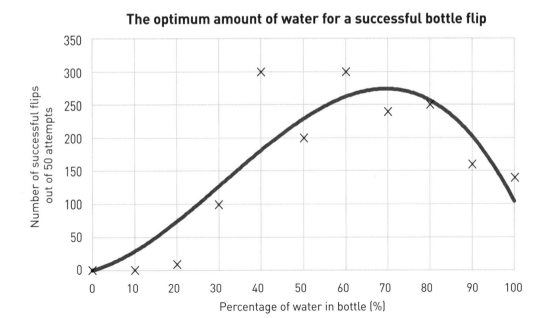

The optimum amount of water for a successful bottle flip

4. Flip my bottle

Sample Discussion

The experiment shows that if there is more water than less, the chances of a successful bottle flip increase. The ideal amount of water is around 60–80% full.

One of the problems with this experiment is that it is not totally fair. The more you flip, the better you can get. This may also explain why there are more successful flips as the experiment progresses. It would be great if you could make a robot to flip the bottle! This would make for a fairer experiment.

Sample Conclusion

The optimum percentage of water for a successful bottle flip is 67%.

Now it's your turn!

Your Results

Amount of water (ml)	Percentage of water (%) (To calculate: Amount of water ÷ volume of bottle x 100	Tally of successful flips

Your Graph

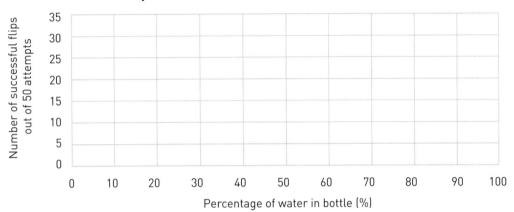

The optimum amount of water for a successful bottle flip

Number of successful flips out of 50 attempts (y-axis): 0, 5, 10, 15, 20, 25, 30, 35

Percentage of water in bottle (%) (x-axis): 0, 10, 20, 30, 40, 50, 60, 70, 80, 90, 100

Your Discussion

Identify the clear trends_____

_____

Discuss any unexpected results_____

_____

Suggest how the experiment could be improved_____

_____

What are the implications of your results to real life applications?_____

_____

Your Conclusion

_____

_____

Further Experiments

Different types of bottles – what is the optimum bottle for a successful bottle flip?

Different size bottles – what is the optimum size bottle for a successful bottle flip?

Do different types of liquid (e.g. vegetable oil or thick paint) increase the number of successful flips?

5. The amazing colour race
Temperature and diffusion

Summary

Get your race gear on people! Hop in and buckle up, we're racing food dye! If you put a drop of food colouring into a glass it slowly starts spreading throughout the water. This is called diffusion. An interesting question to find the answer to is whether the temperature of the water affects how quickly the food colour diffuses.

TIME REQUIRED – 3 hours
DIFFICULTY – 2 out of 5
DANGERS – No obvious dangers

Materials and Equipment
- Rectangular dish
- Water
- Ice
- Food colouring
- Thermometer
- Kettle
- Stopwatch

Question

To determine whether the temperature of water affects the rate of diffusion.

Research

What is diffusion? _____

_____

_____

What is temperature? _____

_____

_____

How does temperature affect diffusion? _____

_____

Hypothesis

The hotter the water the faster/slower the diffusion.

Procedure

1. Fill a rectangular dish with water and add some ice.
2. Once the water has cooled down, remove the ice and measure the temperature.
3. Allow a minute or two for any movement in the water to come to a stop.
4. Squirt a few drops of red food colouring at one end of the dish and time how long it takes the red colour to reach the other end of the dish. Repeat using yellow, then green and finally blue food colouring.
5. Empty the dish and repeat steps 3 and 4 for a range of temperatures.

Variables

Controlled variable	Controlled variable	Controlled variable
The amount of water in the dish	The actual dish used	The type of food colouring used
Controlled variable	**Independent variable**	**Controlled variable**
The amount of food colour used	The temperature of the water (10°C, 20°C, 30°C, 40°C, 50°C)	That the water is perfectly still before adding food colour
Controlled variable	**Controlled variable**	**Controlled variable**
_____ _____	_____ _____	_____ _____

Dependent variable – How long it takes for the food colour to diffuse.

Sample Results

Temperature (°C)	Time for diffusion (seconds)			
	Red	Yellow	Green	Blue
10	360	203	308	233
20	241	200	300	220
30	33	108	110	182
40	17	100	49	92
50	15	20	38	88

Sample Discussion

The general trend is that the hotter the water, the faster the diffusion occurs. The red food colour starts off taking the longest to diffuse at the lowest temperature, but strangely diffuses the most quickly at the highest temperature. This would be great to investigate further.

Sample Graph

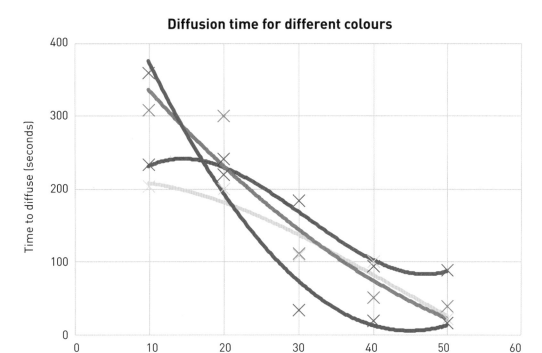

Diffusion time for different colours

Sample Conclusion

The higher the temperature of the water the faster the diffusion occurs.

Now it's your turn!

Your Results

Temperature (°C)	Time for diffusion (seconds)			

Your Graph

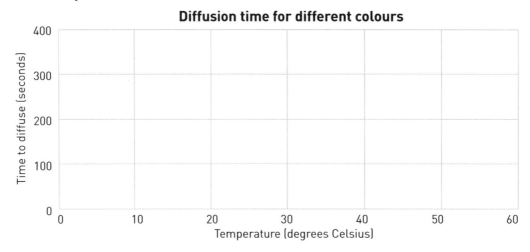

Your Discussion

Identify the clear trends_____

_____

Discuss any unexpected results_____

_____

Suggest how the experiment could be improved_____

_____

What are the implications of your results to real life applications?_____

_____

Your Conclusion

_____

_____

Further Experiments

Speed of diffusion through different types of paper.

Speed of diffusion for different colours through paper.

6. Party sparklers

Do crushed up birthday party sparklers burn faster?

TIME REQUIRED – 2 hours

DIFFICULTY – 2 out of 5

DANGERS – Sparkler dust is very flammable and can burn rapidly. Perform the experiment outside and with adult supervision.

Materials and Equipment

- Two packs of ten party sparklers
- Ceramic tile or old plate
- Matches
- Stopwatch
- Mortar and pestle

Summary

Happy birthday to me! Everyone loves a birthday cake with sparklers. As they slowly burn down towards the cake they throw off tiny little sparkles of happiness. Do you think that there might be a faster way to make a sparkler burn? What if the grey sparkler chemical was broken off and put in a pile? How long would that take to burn up? What about if you used a mortar and pestle and crushed it to a fine powder, would that make it go even quicker? I can feel an experiment coming on!

Question

How does the amount of crushing affect how quickly a pile of sparkler dust burns up?

Research

What is a chemical reaction? _____

_____

When an object is broken up, what happens to its surface area? _____

_____

How does surface area affect the rate of reaction? _____

_____

Hypothesis

The smaller the particles a sparkler is crushed up into, the slower/no change/faster it will burn up.

Circle your best guess

Procedure

1. Take three sparklers and gently break the grey chemical off the wire in as large chunks as possible.

2. Make a pile, shaped like a small mountain, of this chunky sparkler crumble on a ceramic tile.

3. Using another sparkler as a fuse, ignite the pile at the bottom. Record the time it takes for the pile of crushed sparklers to burn up.

4. Taking three more sparklers, remove the chemical, but use the mortar and pestle to lightly crush the chemical into smaller chunks then the previous test. Again, make a pile and time how long it takes to burn up. Record this time.

5. Finally, try crushing three sparklers into a fine powder and timing how long they take to burn up.

Variables

Controlled variable	Controlled variable	Controlled variable
The amount of sparkler chemical	3 sparklers	The type of sparkler used (Acme brand)
Controlled variable	**Independent variable**	**Controlled variable**
The shape of the pile	The size of the sparkler dust particles	Where the pile is ignited from
Controlled variable	**Controlled variable**	**Controlled variable**
_____ _____	_____ _____	_____ _____

Dependent Variable – The time for the sparkler dust to burn up.

Sample Results

Size of particles (3 = large 1 = small)	Time to burn up (seconds)
3 (chunks)	15
2 (small chunks)	10
1 (fine powder)	5

Sample Graph

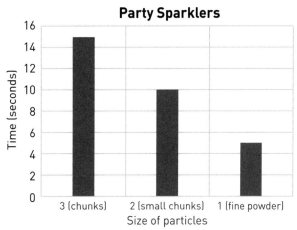

Sample Discussion

As the sparklers were crushed into finer particles, they burnt up more quickly. This is because there is more surface area available for the chemical reaction.

Sample Conclusion

Chemical reaction rate is increased with increased surface area of the reactants.

Now it's your turn!

Your Results

Note: it is always best to repeat tests when doing an experiment in order to improve the reliability of your experiment. There is an extra column to add your results for a second trial.

Size of particles (5 = large 1 = small)	Time to burn up (seconds)		
	Attempt 1	Attempt 2	Average
5 (chunks)			
1 (powder)			

Your Graph

Party Sparklers

Size of particles

Your Discussion

_____

_____

_____

_____

Your Conclusion

_____

_____

Further Experiments

Will crushing up tablets affect how quickly they dissolve in water?

7. Melting coloured ice cubes

Which colour melts quickest in the sun?

TIME REQUIRED – 2 hours (plus over-night for the ice cubes to freeze)

DIFFICULTY – 2 out of 5

DANGERS – No obvious dangers

Materials and Equipment

- Ice cube tray
- Water
- Food dye colours (red, blue, yellow and green)
- A freezer
- A tray
- Stopwatch

Summary

Most people know that a dark-coloured car gets hotter than a light-coloured car. In this fun and colourful experiment we can find out which colours absorb the most heat.

Question

Does the colour of an ice cube affect how quickly it melts?

Research

What are the colours of the rainbow? _____

What causes the colour of light? _____

_____

Why do different objects appear the colour they do in white light?
(You'll need to include a diagram as part of your answer)

_____

_____

_____

Hypothesis

I think the _____ (colour)
ice cube will melt the quickest.

Put your diagram in the box

Procedure

1. Fill an ice cube tray with water.

2. Leave the first two cubes as plain water.

3. Add red food colouring to two of the cubes and stir with the end of a spoon.

4. Now make two blue cubes, two yellow cubes and two green cubes. Make two black cubes by adding all the food colours together.

5. Put the ice cube tray in the freezer and wait until the water has frozen.

6. Tip the cubes out and put them out in the sun on a tray.

7. Time how long it takes for each ice cube to fully melt. Record these times in the table.

(A great idea would be to timelapse photograph the experiment as well.)

Variables

Controlled variable	Controlled variable	Controlled variable
The size of the ice cubes	The ice cubes in the same position in the sun	The same time of day and sun conditions
Controlled variable	Independent variable	Controlled variable
The starting temperature of the ice cubes	The colour of the ice cubes	The surface the cubes are resting on
Controlled variable	Controlled variable	Controlled variable
_____ _____	_____ _____	_____ _____

Dependent Variable –The time it takes for the ice cubes to melt.

Control

This experiment has what is called a control. In this case the control is the ice cubes without any colour added.

Sample Results

Colour of ice cubes	Time to melt (minutes)		
	Cube 1	Cube 2	Average
No colour	70	74	72
Red	56	56	56
Blue	51	53	52
Yellow	61	63	62
Green	55	55	55
Black	54	54	54

Sample Graph

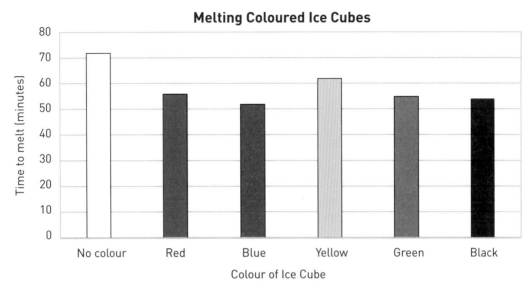

Melting Coloured Ice Cubes

(Bar graph: y-axis "Time to melt (minutes)" from 0 to 80; x-axis "Colour of Ice Cube" with categories No colour, Red, Blue, Yellow, Green, Black)

No colour ≈ 72, Red ≈ 56, Blue ≈ 52, Yellow ≈ 62, Green ≈ 55, Black ≈ 54

Sample Discussion

Surprisingly it was the blue ice cube that melted the quickest and not the black ice cube. The ice cube with no colour took the longest to melt. This could mean a dark-coloured car would get hotter in the sun then a lighter-coloured car.

Sample Conclusion

This experiment showed the darker the colour of the ice cube the quicker it melts.

Your Results

Now it's your turn!

Colour of ice cubes	Time to melt (minutes)		
	Cube 1	Cube 2	Average
No colour			

Your Graph

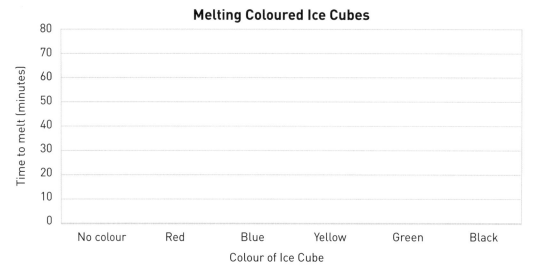

Melting Coloured Ice Cubes

Your Discussion

_____

_____

_____

_____

Your Conclusion

_____

_____

Further Experiment

How does changing the colour of a liquid affect how quickly it heats up in the sun?

How does changing the surface (e.g. metal, plastic, ceramic) the ice cubes are sitting on affect how quickly they melt?

8. Balloon racers
Does size matter?

TIME REQUIRED – 2 hours
DIFFICULTY – 3 out of 5
DANGERS – No obvious dangers

Summary

Everybody loves rockets! What could be more fun than having a rocket race across your living room? A taut string between two walls guides your balloon rocket across the room. But what is the perfect amount of air for the fastest rocket? As the balloon pushes the air out backwards, the air pushes the balloon forwards – a classic case of Newton's Third Law in action.

Materials and Equipment
- 10 metres of string
- 5 drinking straws
- Sticky tape
- 5 balloons
- Ruler
- Stopwatch

Question

What is the optimum amount of air for the fastest balloon rocket?

Research

What is Newton's Second Law?

_____

_____

Apply Newton's Second Law to a rocket _____

_____

_____

Explain how a rocket can work in space. You'll need to apply Newton's Third Law *and* use a diagram __

_____

_____

Hypothesis

The optimum amount of air for a balloon rocket is a little bit/a medium amount/the maximum amount.

Put your diagram in the box

Procedure

1. Thread a long length of string through a plastic drinking straw.
2. Tie the length of string between two points in the house about 8 metres apart.
3. Use sticky tape to attach a balloon to the straw.
4. Blow some air into the balloon so that it partly fills. Measure the diameter of the balloon and record it in the table.
5. Release the balloon at one end of the room and time how long it takes to travel a particular distance. Record the time and distance in order to calculate the average speed (Average speed = distance ÷ time).
6. Repeat step 4 and 5 a total of three times.
7. Now try different amounts of air in the balloon in order to measure distances travelled and the time taken.

Variables

Dependent Variable – Average speed of balloon racer.

Sample Results

(Note: Speed = Distance travelled ÷ Time)

Controlled variable	Controlled variable	Controlled variable
The type of balloon	The shape of the balloon	The colour of the balloon
Controlled variable	**Independent variable**	**Controlled variable**
The length of straw on the balloon	The size the balloon is inflated to.	The tautness of the string across the room
Controlled variable	**Controlled variable**	**Controlled variable**
The type of string	_____ _____	_____ _____

Diameter of balloon (cm)	Trial	Distance travelled (m)	Time(s)	Speed (m/s)	Average Speed (m/s)
10	1	7	2.2	3.2	3.2
	2	6	2	3	
	3	8	2.3	3.5	
20	1	8.5	2.4	3.5	3.5
	2	9	2.5	3.6	
	3	9	2.6	3.5	
30	1	10	2.8	3.6	3.7
	2	10	2.7	3.7	
	3	10	2.6	3.8	

Sample Graph

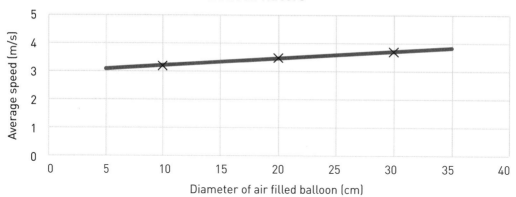

Balloon Racers

Sample Discussion

The average speeds for the three different amounts of inflation were quite similar. The larger balloons started off a bit slower (probably due to their mass) but seemed to have a faster finishing speed. The larger balloons also travelled a bit further as they had more air to propel them.

Sample Conclusion

The more air you blow into a balloon the further, and faster it travels.

Now it's your turn!

Your Results

Diameter of balloon (cm)	Trial	Distance travelled (m)	Time(s)	Speed (m/s)	Average Speed (m/s)
10	1				
	2				
	3				
20	1				
	2				
	3				
30	1				
	2				
	3				

Your Graph

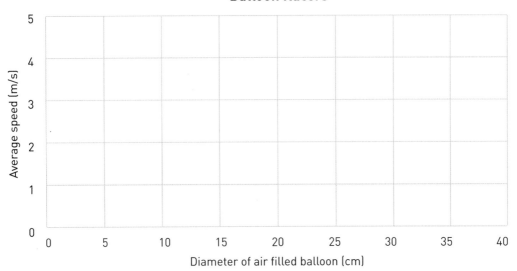

Balloon Racers

Average speed (m/s) vs Diameter of air filled balloon (cm)

Your Discussion

_____

_____

_____

_____

_____

Your Conclusion

_____

_____

Further Experiment

Does the shape of a balloon affect how fast it travels?

Does the colour of a balloon affect how fast it goes?

9. Popping bags!
How to make the best pop!

What's more acidic? Vinegar or lemon juice?

Summary

Plastic popping bags are a fun way to prank your little brother or sister. When the plastic bag pops it's really loud and sprays liquid everywhere!

Question

What are the optimum ingredients for a plastic bag 'popper'?

Research

What is the chemical reaction called when an acid reacts with a base?

_____

What is produced when vinegar reacts with bicarbonate of soda? _____

How does the gas 'pop' open the balloon? _____

_____

Hypothesis

Vinegar will be a more/less effective acid to use than lemon juice.

Procedure

1. Fill a small plastic zip-lock bag with vinegar. Make sure the bag is so full that you can only just close it.
2. Put 3 big tablespoons of bicarbonate of soda into the large plastic zip-lock bag.
3. Put the small zip-lock bag containing the vinegar into the large zip-lock bag and then seal the large bag.
4. To explode the bag, you need to push on the small zip-lock bak so that it bursts open, allowing the vinegar to mix with the bicarbonate of soda.

TIME REQUIRED – 2 hours

DIFFICULTY – 3 out of 5

DANGERS – No obvious dangers. Do not use any other types of acids other than vinegar or lemon juice or injury may occur.

Materials and Equipment

- Small plastic zip-lock bag
- Large plastic zip-lock plastic bag
- Box of bicarbonate of soda (baking soda)
- Vinegar
- Stopwatch
- Lemon juice

5. Time how long it takes for the resulting gas to '*pop*' open and explode the large zip-lock bag.

6. Now repeat the experiment, but this time use lemon juice instead of vinegar.

Variables

Dependent Variable – The time it takes for the bag to pop open.

Controlled variable	Controlled variable	Controlled variable
The amount of liquid acid used (full, small zip-lock bag)	The amount of bicarbonate of soda used (3 full tablespoons)	The type of zip-lock bags used
Controlled variable	**Independent variable**	**Controlled variable**
The size of the zip-lock bags	The acid used in the small bag (vinegar or lemon juice)	The temperature of the liquid acids
Controlled variable	**Controlled variable**	**Controlled variable**
_____ _____	_____ _____	_____ _____

Your Results

Type of acid used	Time to pop (seconds)		
	Time 1	Time 2	Average
Vinegar			
Lemon juice			

Your Graph

This will be a column graph.

Your Discussion

_____

_____

_____

_____

Your Conclusion

_____

_____

10. Cabbage juice indicator

Which citrus fruit is the most acidic?

TIME REQUIRED – 2 hours

DIFFICULTY – 3 out of 5

DANGERS – Take care with the stove and hot water

Materials and Equipment

- Purple cabbage
- Water
- Saucepan
- Stove
- Measuring jug
- Bicarbonate of soda (baking soda)
- Orange, lemon, lime, mandarin or any other citrus fruit

Summary

Purple cabbage is a substance that changes colour in the presence of an acid or a base. When boiled the water goes purply red. When a base such as bicarbonate of soda is added it changes colour to _____. To bring it back to the original colour, an acid such as lemon juice needs to be added.

Question

Which citrus fruit is the most acidic?

Research

List some properties of acids.

_____

_____

_____

What is an indicator? _____

List some natural indicators.

_____

_____

_____

What is neutralisation? _____

Hypothesis

I think lemon/lime/orange/mandarin/grapefruit will be the most acidic.

Procedure

1. Cut up ¼ of a purple cabbage and boil it in two cups of water in a saucepan for two minutes.

2. Pour off and keep the purple liquid. This is your *neutral* indicator solution.

3. Use a measuring jug to put 100 ml of purple cabbage juice into a drinking glass.

4. Add a teaspoon of bicarbonate of soda to the glass and stir in. The cabbage juice should change colour as it is now basic.

5. Use a measuring jug to slowly pour in some orange juice while stirring. Stop adding the orange juice when the cabbage juice has changed back to its original colour. You have now neutralised the bicarbonate of soda with the orange juice. Record how many millimetres of orange juice was required.

6. Repeat steps 3–5 but this time test a variety of citrus fruits, one at a time of course! Make sure you start with fresh cabbage juice each time.

Variables

Controlled variable	Controlled variable	Controlled variable
The amount of bicarbonate of soda used (one teaspoon)	The amount of cabbage juice used to start with. (100 ml)	The temperature of the cabbage juice
Controlled variable	**Independent variable**	**Controlled variable**
_____ _____	The type of citrus fruit used (orange, lemon, lime, mandarin, grapefruit)	_____ _____
Controlled variable	**Controlled variable**	**Controlled variable**
_____ _____	_____ _____	_____ _____

Dependent Variable – The amount of the citrus fruit juice used.

Sample Results

Citrus Fruit	Volume Required (ml)
Orange	55
Lemon	10
Lime	13
Mandarin	60
Grapefruit	20

Sample Graph

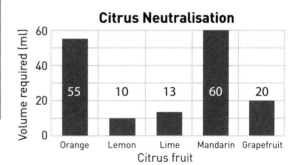

Sample Discussion

Not surprisingly, of the different juices it took the least amount of lemon juice to neutralise the 1 teaspoon of bicarbonate of soda. This shows that lemons are the most acidic. The least acidic was the mandarin, closely followed by the orange.

Sample Conclusion

Lemons are the most acidic citrus fruit.

Your Results

Citrus fruit	Volume required (ml)

Your Graph

Citrus Neutralisation

Your Discussion

_____

_____

_____

Your Conclusion

_____

_____

Further Experiment

Acidity level of different fruits. What kind of oranges are the most acidic?

11. Oobleck
The best recipe ever!

TIME REQUIRED – 2 hours
DIFFICULTY – 2 out of 5
DANGERS – No obvious dangers

Materials and Equipment
- Measuring jug
- Cornflour (cornstarch)
- Measuring jug
- Food colouring
- Water
- A big bowl

Summary

Every kid loves playing with oobleck! It's gooey, messy and just plain weird! Oobleck or magic mud as it's sometimes called, has some very unusual physical properties. Scientifically it's known as a non-Newtonian liquid, but my kids just call it goop! In this experiment we investigate the ideal recipe.

Question

What is the ideal recipe for oobleck? Or more scientifically, how does the ratio of water to cornflour affect the non-Newtonian properties of Oobleck?

Research

What are some examples of non-Newtonian liquids? _____

_____

What are the properties of a non-Newtonian fluid? _____

_____

Hypothesis

Sometimes it's a bit tricky having a hypothesis and this experiment is a case in point! Maybe just a guess of the amount of water with a certain amount of cornflour would be suitable for this experiment. I'm going to guess 100 grams of cornflour and 50 millilitres of water will make the best oobleck.

Procedure

1. Measure 100 g of cornflour into a bowl. Run

your fingers through it and make detailed observations.

2. Add your favourite colour of food colouring to 100 ml of water.

3. Add 20 ml of the coloured water to the cornflour and mix it in with a spoon and your fingers. Again, make detailed observations of how it feels and how it moves when a force is exerted.

4. Add another 20 ml of water and mix it again. How does the mixture respond when it is tipped into another bowl? What about if you tap it with your finger or the blunt end of a pencil. Record your observations.

5. Continue to add 20 ml of water at a time until the mixture gets closer and closer to the desired consistency. Make and record observations every time water is added.

6. Find out what happens if you add more and more water.

Variables

Controlled variable	Controlled variable	Controlled variable
Starting amount of cornflour (100 grams)	The type of cornflour used	The temperature of the water added
Controlled variable	**Independent variable**	**Controlled variable**
The type of bowl used	The amount of water added	_____ _____
Controlled variable	**Controlled variable**	**Controlled variable**
_____ _____	_____ _____	_____ _____

Dependent Variable – The non-Newtonian properties of the mixture.

Sample Results

Water added (ml)	Observations – how it feels, flows, responds when poked etc.
0	It's like a strange cooking flour, not as soft. It's a powder, but it feels strangely hard, like the particles are packed together.
20	Parts of the cornflour have gone hard and clumpy. It's mostly still powder though.
40	The cornflour is becoming more liquidly and a little gloopy in parts.
60	After mixing well, it now seems like the perfect oobleck! It is runny when I slowly poke it with my finger, but feels quite hard when I poke it quickly.

Sample Discussion

Oobleck (or magic mud as some call it) is so very strange! I can make a ball from it if I move it quickly, but the moment I stop moving it, the oobleck runs through my fingers! It is difficult to say what the perfect recipe is because some people might like it runnier or harder than other people.

Sample Conclusion

Using 80 millimetres of water and 100 grams of cornflour makes a very non-Newtonian mixture!

Your Results

Water added (ml)	Observations – how it feels, flows, responds when poked etc.

Now it's your turn!

Your Discussion

_____

_____

_____

_____

Your Conclusion

_____

_____

Further Experiment

Does cornflour form oobleck with any other liquids? For example, milk or vegetable oil.

Does the temperature of the water used affect the properties of the oobleck?

Investigate the optimum recipe for another interesting 'solid' using psyllium husks and water.

12. Oobleck dance

The best sound frequency for making your oobleck dance!

TIME REQUIRED - 3 hours
DIFFICULTY - 4 out of 5
DANGERS - Take care with any sharp tools

Materials and Equipment
- Some oobleck from the previous experiment
- Some basic workshop tools
- A portable speaker
- A smartphone with a tone generator app installed

Summary

Making oobleck dance is even more fun than playing with it using your hands. Do a quick YouTube search of dancing oobleck to see what the fun (and mess) is all about! But, warning, it's not very easy to achieve. The oobleck doesn't always dance and so in this experiment we try and find the best conditions to make it boogie!

Question

What is the best frequency of sound or music to make oobleck dance?

Research

What is oobleck? _____

What is a non-Newtonian fluid? _____

_____

Why does oobleck dance the way it does when put on a loudspeaker?

_____

In terms of soundwaves, what are amplitude and frequency?_____

_____

Hypothesis

The optimum frequency for dancing oobleck will be _____ hertz.

Procedure

1. Using workshop tools carefully remove the protective grill from a portable loudspeaker so as to expose the loudspeaker itself.
2. Download a tone or frequency generator app onto a smartphone (there are a number of free apps available).
3. Put some of your oobleck onto the loudspeaker and play sounds of different frequencies and volumes.

Independent Variable – The frequency of the sound.

Dependent Variable – How the oobleck dances.

Your Results

Frequency of sound (hertz)	Observations – how the oobleck moves, responds or dances

Discussion

_____

_____

Conclusion

_____

_____

Further Experiment

What type of music or songs best make oobleck dance?

Is there an ideal volume to make the oobleck dance?

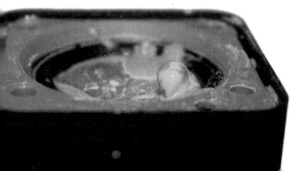
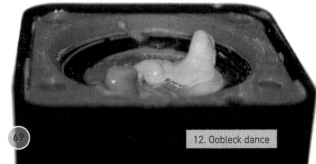

12. Oobleck dance

13. Takeaway coffee cups

The best takeaway coffee cup design for keeping coffee hot

TIME REQUIRED – 3 hours

DIFFICULTY – 3 out of 5

DANGERS – Be very careful with hot water from the kettle

Materials and Equipment
- At least 5 different takeaway coffee cups with different features
- Kettle
- Water
- Thermometer

Summary

Many adults love a hot cup of coffee in the morning. There are so many different types of takeaway coffee cups. Wouldn't it be interesting to find out which cups keep the coffee the hottest for the longest time?

Question

How do the design features of a takeaway coffee cup affect how well it keeps the heat in?

Research

What is heat? _____

_____

What is temperature? _____

_____

Discuss Newton's Law of Cooling _____

_____

_____

_____

How is heat transferred by conduction? _____

_____

_____

Hypothesis

This hypothesis is a little tricky. A good hypothesis will probably focus on a feature of a cup, not on where it was purchased. I think a suitable hypothesis might be 'The takeaway coffee cup that keeps the coffee the hottest will be the one with the thickest sides.'

Procedure

1. Fill each of the different types of takeaway coffee cups with boiling water from a kettle and record the initial temperature.

2. Every two minutes, for half an hour, measure and record the temperature of the water in each of the cups.

3. Once finished, dissect the cups to determine the key features of their heat insulating properties.

13. Takeaway coffee cups

Variables

Controlled variable	Controlled variable	Controlled variable
Starting temperature of the water	The same liquid in each of the cups (water)	Each cup must have a similar volume/capacity
Controlled variable	**Independent variable**	**Controlled variable**
The room temperature must be the same if doing this experiment at different times	The takeaway coffee cup	The cups must all have their lids off or lids on
Controlled variable	**Controlled variable**	**Controlled variable**
The cups must be all standing on the same surface	_____ _____	_____ _____

Dependent Variable – The temperature of the water in the cup.

Results

Time (min)	Temperature (degrees Celsius)				
	Cup 1	Cup 2	Cup 3	Cup 4	Cup 5
0	100	100	100	100	100
2					
4					
6					
8					
10					
12					
14					
16					
18					
20					

Graph

Temperature versus Time

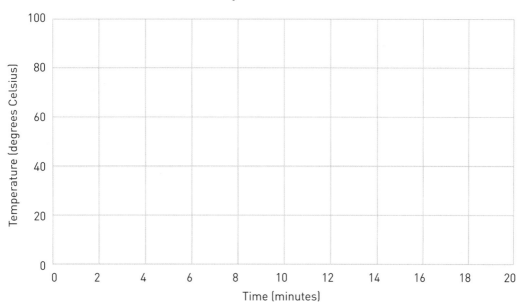

Key for graph

Coffee cup	Colour of line

Coffee Cup Dissection!

Before writing your discussion you should pull the coffee cups apart and see what gives them their heat insulating properties. Is there something special about the cup that stayed hottest for longest? What about the cup that cooled down the quickest – are there features missing that the other cups have?

Discussion

_____

_____

Conclusion

_____

_____

Further Experiment

How important is the use of a lid in keeping coffee hot in a takeaway cup?

What modifications can be made to a standard takeaway coffee in order to make it an even better insulator?

14. Joe's cup of tea

Should Joe add the milk now or later?

TIME REQUIRED – 2 hours

DIFFICULTY – 3 out of 5

DANGERS – Be very careful with hot water from the kettle

Materials and Equipment

- Kettle
- Water
- 2 mugs
- Milk
- Measuring jug
- Thermometer

Summary

This is an unusual science experiment. It really just sets out to solve a problem, but there is some interesting science to be learned along the way.

Joe loves his cup of tea. He enjoys it with milk, but he also likes to drink his tea as hot as possible. One day, he has just finished pouring the boiling water into the mug and as he is about to add the cold milk, the telephone rings. He knows it's his mum calling and she likes to speak for at least 15 minutes. Should Joe pour the milk into the tea before he answers the telephone, or should he wait until after the telephone conversation to pour the milk in? Either way, he knows he can't drink his tea while talking to his mum, and don't forget...he likes his tea as hot as possible!

Question

When should Joe pour the milk into his tea? Before the answering the telephone or after the telephone conversation?

Research

What is heat?_____

_____

What is temperature? _____

_____

Discuss Newton's Law of Cooling _____

_____

_____

How is heat transferred by convection? _____

_____

Hypothesis

Joe should pour the milk in before/after the telephone call.

Procedure

1. Boil the kettle and nearly fill two mugs with hot water.
2. Pour 40 ml of cold milk into one of the mugs and stir it in.
3. Measure and record the temperature of the liquids in both mugs.
4. Keep measuring and recording the temperature every minute for 15 minutes.
5. At the end of 15 minutes, pour 40 ml of cold milk into the second mug, stir it in and measure and record the temperature.

Variables

Controlled variable	Controlled variable	Controlled variable
The amount of water to start with	The starting temperature of the water	The starting temperature of the milk
Controlled variable	Independent variable	Controlled variable
The amount of milk added	When the milk is added	The same cups used
Controlled variable	Controlled variable	Controlled variable
The same room conditions for both cups	_____	_____

Dependent Variable The final temperature of the cup of tea.

Results

Time (minutes)	Temperature (degrees Celsius)

	Sample results cup 1	Sample results cup 2	Your results cup 1	Your results cup 2
	Starting temperature	Milk added at start		
	73	86		
1	73	78		
2	70	76		
3	70	76		
4	65	71		
5	64	69		
6	61	68		
7	61	68		
8	60	65		
9	60	65		
10	57	63		
11	56	62		
12	56	61		
13	55	61		
14	54	58		
15	53	Milk added at end 50		

Graph

This will be a line graph. You will need to plot *two* lines on the graph, one line for each cup.

14. Joe's cup of tea

You could plot the sample results, or even better, do the experiment yourself and plot your own results.

Discussion

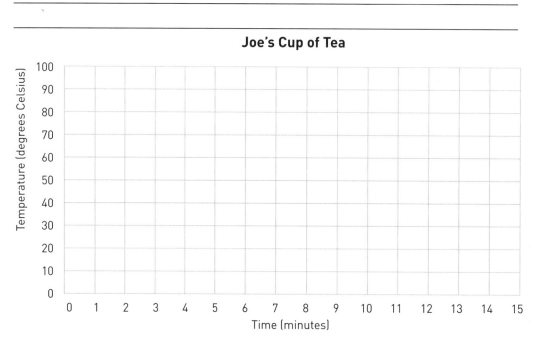

Joe's Cup of Tea

Conclusion

Further Experiment

Which freezes more quickly? A tray of boiling water or a tray of cold water put in the freezer at the same time? (This experiment, and the results, has caused a lot of controversy and confusion in the scientific community!)

15. An eggciting eggsperience

Free range vs Caged eggs bounce off

TIME REQUIRED – 2 hours (excluding 2 days for eggshells to dissolve)
DIFFICULTY – 4 out of 5
DANGERS – No obvious dangers

Materials and Equipment
- 6 free-range eggs
- 6 caged eggs
- 2 litres white vinegar
- 2 large bowls
- Smartphone or video camera
- Ruler

Summary

When vinegar is used to dissolve the shell of an egg you're left with this wonderful bouncy, squishy, eggy ball! The outer rubbery skin of the egg under the shell is called the membrane. An interesting question is whether the strength of the membrane is the same or different for different types of eggs.

Question

Does the strength of an egg's membrane depend on whether the egg is from a free-range chicken or a caged chicken?

Research

What is a chicken egg?_____

_____

What is a free-range chicken? _____

_____

What is a caged chicken?_____

_____

What is the membrane of egg composed of? _____

Hypothesis

I think the membrane strength of a free-range/caged chicken egg will be the strongest.

Procedure

1. Put three free-range eggs in a bowl. Pour in white vinegar until the eggs are fully submerged.
2. Put three caged eggs in a bowl. Pour in white vinegar until the eggs are fully submerged.
3. Allow to stand for two days while the eggshells dissolve. You may need to carefully rub some of the shells off with your fingers after a day or so.
4. Drop one of the free-range eggs from 6 cm high onto a hard surface. Use the slow-motion feature of a smartphone and a ruler to film how high it bounces. Repeat for a total of three times.
5. Repeat step 4 at increasing increments of 2 cm height. Continue to do so until the membrane breaks.
6. Repeat steps 4 and 5 for the other free-range egg.
7. Finally repeat steps 4–6 for the caged eggs.

(Note: There is a spare of each egg in case something goes wrong!)

Variables

Controlled variable	Controlled variable	Controlled variable
Heights from which eggs are dropped	Surface the egg is dropped onto	_____ _____
Controlled variable	**Independent variable**	**Controlled variable**
_____ _____	Free range or caged eggs	_____ _____
Controlled variable	**Controlled variable**	**Controlled variable**
_____ _____	_____ _____	_____ _____

Dependent Variable – The height of the bounce.

Results

Free Range Eggs

Bounce heights	Height from which egg is dropped (cm)																							
	6				8				10				12				14				16			
Trial	1	2	3	Av	1	2	3	Av	1	2	3	Av	1	2	3	Av	1	2	3	Av	1	2	3	Av
Egg 1																								
Egg 2																								
Average																								

Caged Eggs

Bounce heights	Height from which egg is dropped (cm)																							
	6				8				10				12				14				16			
Trial	1	2	3	Av	1	2	3	Av	1	2	3	Av	1	2	3	Av	1	2	3	Av	1	2	3	Av
Egg 1																								
Egg 2																								
Average																								

15. An eggciting eggsperience

Graph

This will be a line graph. You will need to plot *two* lines on the graph, one line for the free range eggs and one line for the caged eggs. Plot the average bounce height (shaded) versus the height from which the egg is dropped.

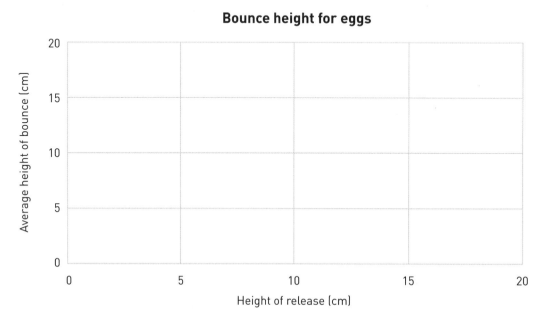

Bounce height for eggs

Discussion

_____

_____

_____

_____

Conclusion

_____

_____

Further Experiment

How does the strength of the membrane of a duck egg compare to a chicken egg?

16. Great balls of fire!

Burning powders

TIME REQUIRED – 3 hours

DIFFICULTY – 4 out of 5

DANGERS – This experiment involves fire and so is high risk. It must be done with adult supervision and outside, away from any flammable materials.

Summary

Believe it or not but cooking flour is actually quite flammable. If you try to burn a spoonful of it nothing much happens. But if you blow the flour through a flame it bursts into a great ball of fire!

Question

Which powdered foods are the most flammable?

Materials and Equipment

- 50 cm garden hose
- Funnel
- A portable gas camp stove
- Measuring tape
- Video camera or smart phone
- Plain flour, self-raising flour, cornflour, cocoa powder, icing sugar, other food powders you may have in the kitchen cupboard

Research

What is a chemical reaction? _____

_____

_____

Describe the chemical reaction of combustion or substances burning _____

_____

_____

How does the surface area affect the rate of a chemical reaction? _____

_____

What is the molecular structure of flour? _____

What is the molecular structure of sugar? _____

Hypothesis

The most flammable foodstuff powder in the kitchen is _____

Procedure

1. Attach a 50 cm length of garden hose to a kitchen funnel.
2. Set up a gas stove outside away from any flammable materials.
3. Set up a measuring tape in the background and a video camera to film the chemical reaction.
4. Put half a cup of plain flour into the funnel.
5. Blow the flour through the flame of a portable gas stove. Film the resulting ball of fire. Repeat.
6. Repeat with each of the powdered foodstuffs.
7. Review the footage, make observations and attempt to gauge the size of each of the balls of flame.

Variables

Controlled variable	Controlled variable	Controlled variable
The amount of powder	How hard you blow on the hose	The angle you hold the funnel to the flame
Controlled variable	**Independent variable**	**Controlled variable**
The size of the flame	The type of powder	
Controlled variable	**Controlled variable**	**Controlled variable**
The strength of the wind outside	_____ _____	_____ _____

Dependent Variable – The flammability of the powder.

Results

The powder	Observations	Diameter of ball of flame (cm)		
		1	2	Avg
Plain flour				
Self-raising flour				
Cornflour				
Chocolate powder				
Cocoa powder				
Icing sugar				

16. Great balls of fire!

Graph

This will be a column graph.

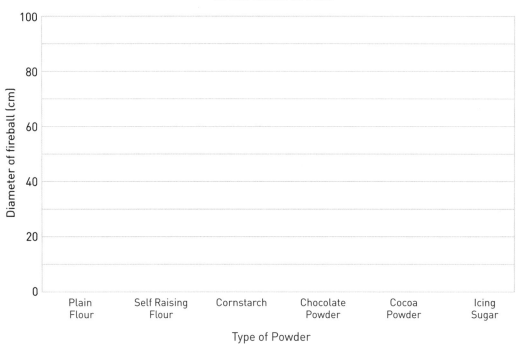

Great Balls of Fire

Discussion

_____

_____

Conclusion

_____

_____

Further Experiments

How does the amount of powder in the funnel affect the size of the ball of flame?

How does the size of the funnel affect the size of the ball of flame?

17. Toy-cup flyers
Does size matter?

TIME REQUIRED – 4 hours
DIFFICULTY – 4 out of 5
DANGERS – No obvious dangers

Materials and Equipment
- A range of paper and plastic cups of different diameters
- Masking tape
- Elastic bands
- Measuring tape
- Stopwatch

Summary

Everyone loves paper planes. Well these toy-cup flyers are just as fun to fly, and even more interesting! You can make really small ones and even massive ones from takeaway chicken buckets! The big question is, what's the best size for the longest flight?

Question

How does changing the size of the cups of a toy-cup flyer affect how well it flies?

Research

What is the Bernoulli effect? _____

_____

What is the Magnus effect? _____

_____

Where is the Magnus effect used? _____

How does a toy-cup flyer actually fly? _____

_____

Hypothesis

I think the best toy-cup flyers will be those made from small/medium/large cups.

Procedure

1. Make a toy-cup flyer by joining two small,

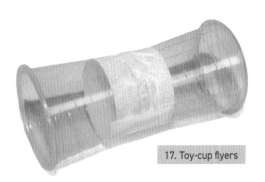

equal-sized cups together at their bases using masking tape.

2. Join a number of elastic bands together to make one long one.

3. Wind the giant elastic band tightly around the middle of the toy-cup flyer (where the bases of the cups are joined) and then release it.

4. Measure how far it flies and how long it's in the air. Repeat a total of 10 times.

5. Now try building a number of different sized toy-cup fliers and testing them.

Independent Variable – The size of the cups used.

Dependent Variable – How well/far the toy-cup flyer flies.

Results

Biggest diameter of cup (cm)	Length of the cup (cm)	What cup is made from	Distance travelled (cm)	Time in air (s)	Observations

Discussion

Conclusion

Further Experiment

Which fly better, paper cups or plastic cups?

How important is the speed of the spin to a successful flight?

17. Toy-cup flyers

18. Milk rocks!
Protein content of different milks

TIME REQUIRED - 3 hours (plus a few days for drying time)

DIFFICULTY - 3 out of 5

DANGERS - Take care not to overheat the milk

Summary

Little Miss Muffet sat on a tuffet eating her curds and whey! When you warm up milk and add vinegar, the proteins coagulate forming curds. This solid can then be filtered from the mixture and used to make cheese. With the large variety of milks in the supermarket, an interesting question is which type produces the most curds?

Materials and Equipment

- Large saucepan or bowl
- A variety of different types of milks (1 litre of each)
- 5 litres of vinegar
- Stove or a microwave
- Wooden spoon
- Old tea towel
- Kitchen scales

Question

Which milk in the supermarket contains the most protein?

Research

What actually is milk? _____

_____

_____

_____

How does vinegar cause the milk to coagulate?

_____

What are curds and whey? _____

What's the difference between different types of milk? For example, low fat, full cream etc. __

_____

Hypothesis

I think the milk with the most protein will be _____

Procedure

1. In a large bowl, warm up 1 litre of milk to about 35 degrees Celsius (body temperature). You could use a microwave or a stove.

2. Add a litre of white vinegar to the milk and slowly stir with a wooden spoon.

3. The milk should start curdling almost immediately. Once it has finished curdling, pour out and discard the liquid remaining.

4. Put the solids into the tea towel and squeeze out any excess liquid.

5. Make a cube from the solids and leave it to dry in a warm place until it goes hard (this will take a few days).

6. Once the cube has dried, you can weigh it with the kitchen scales. This will give an indication of the amount of protein present in the milk.

7. Complete a similar process with each of the different milks.

Variables

Controlled variable	Controlled variable	Controlled variable
The amount of milk used	The amount of vinegar used	The starting temperature of the milk
Controlled variable	**Independent variable**	**Controlled variable**
The length of time milk coagulates for	The type of milk (full cream, low fat, skim etc.)	How hard the curds and whey mixture is squeezed
Controlled variable	**Controlled variable**	**Controlled variable**
_____ _____	_____ _____	_____ _____

Dependent Variable – The mass of the milk rock produced.

Results

Type of milk	Mass of milk 'rock' formed (grams)	Observations
Full cream		
Low fat		

Graph

This will be a column graph.

Amount of protein in milk

Discussion

_____

_____

_____

_____

Conclusion

_____

_____

Further Experiment

What other acidic substances will cause the milk to curdle? What about lemon juice or orange juice?

19. The leaky can

TIME REQUIRED – 3 hours
DIFFICULTY – 4 out of 5
DANGERS – Take care when drilling holes

Materials and Equipment
- A large tin can (similar to a baby milk formula can)
- Measuring jug
- Electric drill
- Range of twist drill bit sizes
- Water
- Stopwatch

Summary

One way to empty a tin can could be to drill a hole in the bottom of it. If the hole was made twice as wide, would the tank empty twice as quickly, or is there some other relationship between hole size and the rate at which water flows? Try this experiment to find out!

Question

What is the relationship between the size of an outlet hole in a tin can of water and the rate at which the water flows out?

Research

What is pressure in liquids? _____

Explain Poisson's Law _____

_____

Hypothesis

I think that if you double the size of an outlet hole in a water tank, the rate of flow of water will double/triple/ _____ .

Procedure

1. Find the volume of a baby formula tin by finding out how much water it holds using a measuring jug.
2. Drill a 2 mm diameter hole in the base of a large, round baby formula tin.
3. Cover the hole while you fill the tin with water, then time how long it takes for the water to completely flow out. Repeat.

4. Use a 4 mm drill piece to increase the size of the original hole. Fill the tin with water and again find out how long it takes to empty. Repeat.

5. Repeat step 3 using 6 mm, 8 mm, 10 mm and 12 mm drill pieces.

Independent Variable – The size of the hole.

Dependent Variable – The flow rate of the water.

Flow rate
= Volume of
can (ml) ÷
Average time

Results

Diameter of hole (mm)	Time to empty (seconds)			Flow rate (ml/s)
	1	2	Average	
2				
4				
6				
8				
10				
12				

19. The leaky can

Graph

This will be a line graph. Plot the flow rate on the vertical axis and the diameter of the hole on the horizontal axis.

Hole size and water flow

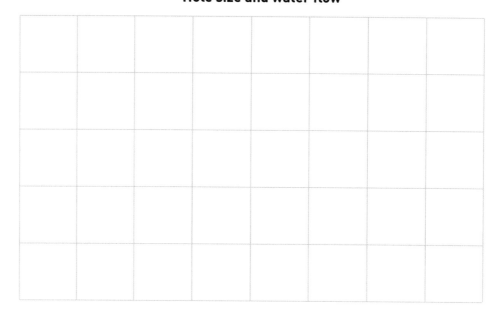

Flow rate

Diameter of the hole (mm)

Discussion

_____

_____

Conclusion

_____

_____

Further Experiment

Try a different liquid such as vegetable oil.

20. Under pressure

Water pressure and flow rate

TIME REQUIRED – 3 hours

DIFFICULTY – 3 out of 5

DANGERS – Don't leave buckets of water around as babies and toddlers can drown in them

Materials and Equipment
- Large, 10-litre bucket
- Plastic hose about 1 metre long
- Water
- Stopwatch

Summary

I love aquariums, fish tanks and ponds. When I was a boy I enjoyed siphoning the water out of them using a length of hose. When the end of the hose was well below the water level, water would siphon quickly, if you raised the end of the hose to the height of the water then the water would stop flowing. How exactly does the height difference affect flow rate?

Question

How does changing the height of a hose affect the flow rate of water?

Research

What is pressure in liquids? _____

_____

How does pressure affect flow rate? _____

_____

Hypothesis

As the height difference is doubled, the water flow doubles/triples/ _____

Procedure

1. Fill the bucket with water and then use a tap to fully fill the hose with water.
2. Put a thumb over each end of the hose so that water cannot escape, then put one end in the bottom of the bucket and lay the other end on the ground.
3. Remove your thumbs and the water should start flowing and emptying the bucket.

20. Under pressure

4. Hold the end of the hose that has water flowing out of it at the same height as the bottom of the bucket and time how long it takes for the water to empty. Repeat.

5. Refill the bucket and place it on a surface above ground level and this time find out how long it takes for it to empty with the hose 10 cm below the bottom of the bucket. Repeat.

6. Repeat step 5 a number of times, each time lowering the hose an additional 10 cm.

Variables

Controlled variable	Controlled variable	Controlled variable
The length of the hose	The diameter of the hose	The liquid (water)
Controlled variable	**Independent variable**	**Controlled variable**
The temperature of the water	The flow rate of the water	The starting amount of water
Controlled variable	**Controlled variable**	**Controlled variable**
_____ _____	_____ _____	_____ _____

Dependent Variable – The distance the end of the hose is below the bottom of the bucket.

Results

Distance hose is below bottom of bucket (cm)	Time to empty (seconds)			Flow rate (ml/s)
	1	2	Average	
0				
10				
20				
30				
40				
50				

Graph

This will be a line graph. Plot the flow rate on the vertical axis and the distance hose is below bottom of bucket on the horizontal axis.

Discussion

_____

_____

_____

_____

Conclusion

_____

_____

Further Experiment

Does the diameter of the bucket affect the flow rate?

21. Paper ninja stars

Is bigger always better?

TIME REQUIRED – 4 hours
DIFFICULTY – 4 out of 5
DANGERS – Paper ninja stars could cause eye injury

Materials and Equipment
- A4 paper or origami paper
- Measuring tape

Summary

What's more fun than a paper plane? A paper ninja star of course! Easy to make, and fun to experiment with. An interesting question to investigate is finding out if there is an optimum size to make these ninja stars so that they fly the furthest.

Question

How does the size of a paper ninja star affect how far it flies?

Research

What is origami? _____

What are the origins of ninja stars?_____

How does a paper plane actually stay in the air? (you'll need to show some force diagrams)

_____

_____

How does a paper ninja star stay in the air? (this may be quite difficult to find information on!)

_____

_____

Hypothesis

I think the optimum size for a ninja star is to start with a 10/15/20/25 cm square piece of paper.

Procedure

1. Make a paper ninja star from a 10 cm by 10 cm piece of paper. (You will need to watch a video on YouTube to see how this is done as it's too complicated to try and explain in this book!)

2. Practise throwing it in order to find the best way to throw it.

3. Throw it 10 times, in the same way, measuring the range (distance travelled) each time.

4. Now make a range of different sizes of ninja stars and using the same motion to throw each time, find the average distances they travel.

Variables

Controlled variable	Controlled variable	Controlled variable
Throwing the paper ninja stars the same way, each time.	The type of paper used	The colour paper used
Controlled variable	**Independent variable**	**Controlled variable**
Wind conditions (preferably do it inside with no wind)	The size of the paper ninja star	
Controlled variable	**Controlled variable**	**Controlled variable**
_____ _____	_____ _____	_____ _____

Dependent Variable – The distance the paper ninja stars flies.

Results

Original size of paper (cm)	Distance paper ninja star travels (metres)										
	1	2	3	4	5	6	7	8	9	10	Avg
10 by 10											
15 by 15											
20 by 20											

Graph

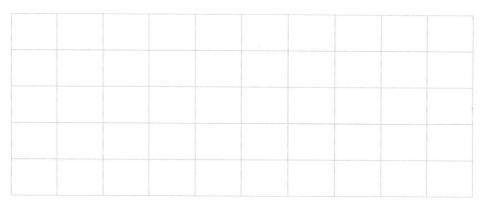

Distance paper ninja star travels (metres)

Discussion

_____

_____

_____

_____

Conclusion

_____

_____

Further Experiment

Does the thickness of the paper used affect the distance of the ninja star?

22. Earth worms and plant growth

Can worms make a difference?

TIME REQUIRED – 5 hours (over 3 weeks of growing time)

DIFFICULTY – 4 out of 5

DANGERS – Take care with dry potting mix as it can be a breathing irritant

Materials and Equipment

- Fine mesh
- 5 large pots
- Large bucket
- Potting mix
- Bucket of partially composted food scraps
- 50 composting worms
- Corn seedlings and bean seedlings

Summary

'Nobody likes me, everybody hates me, I think I'll go eat worms!' No, don't do that! Worms are great for the soil, eat something else instead! Worms help to break down and compost organic matter into soil. As they burrow, they help air and water mix better with the soil as well. I think that they could even possibly help plants grow better. Try this experiment to find out!

Question

Do worms in soil help plants grow better?

Research

Composting worms _____

_____

How do worms improve the soil? _____

_____

Plant growth, what is required? _____

Hypothesis

I think worms in the soil will _____ the growth of plants.

Procedure

1. Push the mesh into the pots so that your worms can't escape through the bottom holes!

2. In a large bucket, mix the potting mixture with partially composted food waste (the food waste is food for the worms).

3. Fill each pot with this new improved potting mix.

4. Leave the first pot without any worms. This is your control.

5. Add 5 worms to the second pot, 10 to the third pot, 15 to the fourth pot and 20 to the fifth pot.

6. Plant some corn and bean seedlings in each of the pots.

7. Water thoroughly and put in a sunny spot.

8. Take a photo every day of the five pots together.

9. After an adequate number of days or weeks, measure the plants.

Variables

Controlled variable	Controlled variable	Controlled variable
The same amount of potting mix	The same amount of added organic material in each pot	The same number of plants in each of the pots
Controlled variable	**Independent variable**	**Controlled variable**
The same type of plants in each of the pots	The number of worms in each pot	The amount of water each pot gets
Controlled variable	**Controlled variable**	**Controlled variable**
The same position in the sun for each of the pots	Same type of worms in the pots	_____ _____

Dependent Variable – The growth of the plants.

Results

Number of worms	Average height of _____ seedlings (cm)								
	Day 0	Day	Day	Day	Day	Day	Day	Day	Day
0									
5									
10									
15									
20									

Number of worms	Average height of _____ seedlings (cm)								
	Day 0	Day	Day	Day	Day	Day	Day	Day	Day
0									
5									
10									
15									
20									

Graph

This will be a line graph. You can plot the average plant heights on the vertical axis and day number on the horizontal axis. One line for each plant.

Average plant heights

Day number

Discussion

_____

_____

_____

_____

Conclusion

_____

_____

Further Experiment

How do different types of worms affect the growth of plants?

23. Seed treatment to promote germination

TIME REQUIRED – 4 hours total (over 2 weeks)

DIFFICULTY – 3 out of 5

DANGERS – Take care with hot water

Materials and Equipment

- 20 bean seeds
- 20 corn seeds
- Warm water
- Potting mix
- Ruler
- 5 plastic pots

Will soaking seeds in water speed up germination?

Summary

Do you have a green thumb? (That means do you like gardening!) If you do, then you'll love this science experiment. Is there a way to make seeds germinate and grow more rapidly? What about soaking the seeds in warm water for different lengths of time? Design and carry out an experiment to find out yourself!

Question

Does pre-soaking seeds affect how quickly they germinate?

Research

What is germination? _____

_____

What pre-treatments are used to promote germination? _____

_____

Hypothesis

_____

23. Seed treatment to promote germination

Procedure

_____

_____

_____

_____

_____

_____

Variables

Controlled variable	Controlled variable	Controlled variable
Controlled variable	**Independent variable**	Controlled variable
Controlled variable	Controlled variable	Controlled variable

Dependent Variable – _____

Results

Graph

Discussion

Conclusion

Further Experiment

What is the ideal temperature of the water to pre-soak a seed in to promote germination?

What is seed scarification and can it promote germination?

24. Vortex rings

Does the size of the hole affect how far the vortex ring travels?

TIME REQUIRED – 4 hours

DIFFICULTY – 4 out of 5

DANGERS – Take care with sharp knives and power tools!

Materials and Equipment
- Electric drill
- Electric jig saw
- A large plastic garbage bin
- Builders plastic to cover the top of bin
- Duct tape
- Smoke machine
- Long measuring tape

Summary

Vortex rings are the most fun ever! I always bring a vortex ring bin to science shows – as I thump the plastic the vortex ring shoots out and whooshes into the faces of kids in the audience. This got me thinking, what's the best size hole for making these vortex smoke rings travel the furthest?

Question

How does the size of the hole affect how far a vortex ring travels?

Research

What is a vortex?_____

_____

Where are vortexes produced naturally? _____

How does a vortex form?_____

_____

Hypothesis

The smaller/larger the hole, the further the smoke vortex ring will travel.

Procedure

1. Using the electric drill and jig saw, cut a 5 cm diameter hole in the base of a large plastic garbage bin.
2. Use the thick builders plastic and the duct tape to cover the top of the bin.

3. Fill the bin with smoke using the fog machine.

4. Thump the plastic with the palm of your hand and measure how far the smoke rings travel.

5. Now increase the size of the hole to 10 cm diameter and see how far the vortex ring travels.

6. Repeat for diameters 15 cm, 20 cm and 25 cm.

Variables

Controlled variable	Controlled variable	Controlled variable
The manner in which you thump the plastic	The amount of fog in the bin	The size of the bin
Controlled variable	**Independent variable**	**Controlled variable**
The tautness of the plastic	The diameter of the hole	The wind conditions (preferably no wind!)
Controlled variable	**Controlled variable**	**Controlled variable**
	_____	_____
	_____	_____

Dependent Variable – The distance the smoke vortex ring travels.

Results

Diameter of hole (cm)	Distance vortex ring travels (metres)					
	1	2	3	4	5	Average
5						
10						
15						
20						

Graph

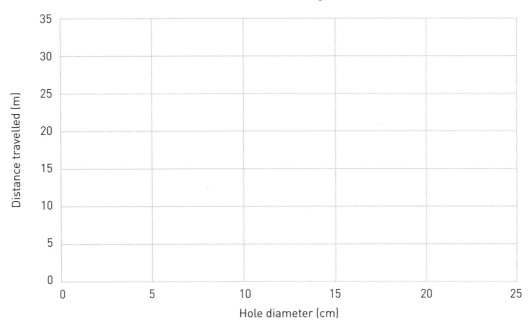

Vortex Rings

Distance travelled (m) vs Hole diameter (cm)

Discussion

_____

_____

_____

_____

Conclusion

_____

_____

Further Experiment

Does the size of the bin affect how far the ring travels?

What is better, to hit the plastic harder or softer for maximum ring travel?

25. Tin can telephones!

How to make the best tin can telephone ever!

TIME REQUIRED – 3 hours
DIFFICULTY – 3 out of 5
DANGERS – Take care with the electric drill

Materials and Equipment

- Electric drill with a thin drill bit
- A variety of empty metal food cans, plastic cups and polystyrene cups
- Different types of string

Summary

No childhood is complete without playing with tin can telephones. When you speak into one of the cans, your voice seems to magically travel along the string so that your friend can hear you in the other can! It's time to experiment and make the best tin can telephone ever!

Question

What are the best type of 'cans' and string to use for a tin can telephone?

Research

What is a compression wave? _____

_____

How does sound travel? _____

_____

How does a tin can telephone actually work? _____

_____

Hypothesis

The best type of 'can' to use for a tin can phone will be a tin can/plastic cup/polystyrene cup.

The best type of string to use for a tin can phone will be nylon/polyester/fishing line/

_____ .

Procedure

1. Use the drill to bore a small hole in the base of two cans.
2. Use 15 metres of string to make a tin can telephone.
3. Test the telephone with a friend, record your observations.
4. Now try different strings and different 'cans', recording your observations each time.

Independent Variable – First the type of 'cans' used, then the type of string used.

Dependent Variable – How well the 'telephone' works.

Results

Type of can _____

25. Tin can telephones!

Type of 'can'	Observations/Comments on how effectively sound is transmitted and received
Tin can	
Plastic cup	
Polystyrene cup	

Results Type of string _____

Type of string	Observations/Comments on how effectively the sound is transmitted and received

Discussion

The best type of 'can' to use was the _____

I think this is because _____

_____

_____

The best type of string to use was the _____

I think this is because _____

_____

_____

Conclusion

To send and receive the best sound, a _____ 'can' should be used to speak into and _____ string should be used to make the connection.

Further Experiment

How does the size of the cans affect the effectiveness of the telephone?

Is it important to have the transmitter can and receiver can the same material, or is it better to have a different arrangement?

26. Food preservation

What preserves food the best?

TIME REQUIRED – 3 hours (over 1 week)

DIFFICULTY – 3 out of 5

DANGERS – Biological hazard. Keep out of reach of small children. Once bacteria and fungi have grown make sure to dispose of carefully and wash glasses thoroughly.

Summary

Food goes bad when natural bacteria and fungi (mainly moulds) grow on it. In this experiment you will investigate food additives that may slow down the rate food goes 'off'.

Question

Which food additive best prevents beef broth from going off?

Materials and Equipment

- Large jug
- Beef stock cube
- Warm water
- 5 drinking glasses
- Spoon
- Salt
- Sugar
- Vinegar
- Lemon juice

Research

What are bacteria and which bacteria cause food to go off? _____

_____

_____

What is mould and which moulds cause food to go bad? _____

_____

_____

What food additives are added to food to prevent it from spoiling? _____

_____

Louis Pasteur and the preventing of food spoilage _____

_____

Hypothesis

I think the best food additive to prevent the beef broth from going off will be _____

Procedure

1. Make up a litre of beef broth in the jug by adding one beef stock cube to a litre of warm water.
2. Pour equal amounts of the broth into the five glasses, making them about 80% full.
3. Leave glass number one as the control.
4. Add a tablespoon of salt into glass two.
5. Add a tablespoon of sugar into glass three.
6. Add a tablespoon of lemon juice into glass four.
7. Add a tablespoon of vinegar into glass five.
8. Label the glasses accordingly.
9. Leave the glasses in an open, warm place for seven days. Take a photograph everyday and put it in the results table.

Variables

Controlled variable	Controlled variable	Controlled variable
_____ _____	_____ _____	_____ _____
Controlled variable	**Independent variable**	**Controlled variable**
_____ _____	_____ _____	_____ _____
Controlled variable	**Controlled variable**	**Controlled variable**
_____ _____	_____ _____	_____ _____

Dependent Variable – _____

Results

Day number	Photograph/drawing/observations of the five glasses
0	
1	
2	
3	
4	
5	
6	
7	

Discussion

Make a comment on each of the additives and the control (no additive) _____

_____

_____

_____

Which additive worked best? Which additive had the least effect? _____

_____

_____

_____

Difficulties, improvements, practical applications _____

_____

_____

_____

Conclusion

_____

_____

Further Experiment

Investigate some other ways of avoiding food spoilage such as heat treatment.

27. Flower preservation

TIME REQUIRED – 4 hours
(over 2 weeks)

DIFFICULTY – 3 out of 5

DANGERS – Careful with any tools

What helps flowers last longer?

Materials and Equipment
- 6 large, plastic soft-drink bottles
- Large bunch of cut flowers
- Sugar
- Salt
- Honey
- Tea and coffee

Summary

I hope you love your mum as much as I do! All mums love getting flowers, so why don't you get her a bunch today? Once you've got them, you want them to last as long as possible. Is it best to put them in just plain water, or is there something you can add to the water to make them last longer?

Question

What can be added to the water in a vase so that the cut flowers last longer?

Research

What are flowers?_____

_____

What nutrients do plants require?_____

_____

What are some methods to help cut flowers last longer? _____

_____

_____

Hypothesis

I think adding _____ to the water will make cut flowers last longer.

Procedure

This experiment is quite straight forward. Set up six bottles of water, with two flowers in each bottle and a different substance (such as sugar) mixed in each bottle. Keep one bottle with just plain water, this is your *control*. Write a procedure in point form below:

1. _____

2. _____

3. _____

4. _____

5. _____

Draw your set up and label it:

Variables

Controlled variable	Controlled variable	Controlled variable
_____ _____	_____ _____	_____ _____
Controlled variable _____ _____	**Independent variable** The substance added to the water in the vase	**Controlled variable** _____ _____
Controlled variable _____ _____	Controlled variable _____ _____	Controlled variable _____ _____

Dependent Variable – The condition of the flowers.

Results

Water treatment	Condition of flowers on each day (Score out of 10 and a comment)				
	1	2	3	4	5
Control – Nothing added to the water					

Water treatment	Condition of flowers on each day (Score out of 10 and a comment)				
	6	7	8	9	10
Control – Nothing added to the water					

Discussion

_____

_____

_____

_____

Conclusion

_____

_____

Further Experiment

Some florists sell a sachet with a flower preservative mixture to add to the water, it would be interesting to test how well this preserves cut flowers.

Some florists suggest lightly crushing the base of the stems to preserve the cut flowers, this would also be interesting to investigate.

27. Flower preservation

28. Ball bearings through oil

Does size matter?

TIME REQUIRED - 3 hours

DIFFICULTY - 4 out of 5

DANGERS - Take care with glass

Materials and Equipment

- A large plastic bottle
- Enough vegetable oil to fill the tube
- Different sized ball bearings 4–10 mm (available online)
- Stopwatch

Summary

This is a great investigation for the person who enjoys mathematics! Ball bearings falling through oil seems simple enough, but there is a twist! The problem is that as you test different sizes of ball bearings falling through the oil, you are also changing the weight as well. This makes life a bit complex and will need some serious thought to analyse the results!

Question

How does changing the size of a ball bearing affect how quickly it falls through vegetable oil?

Research

What is gravity?_____

_____

What is weight? _____

What is drag?_____

Investigate terminal velocity _____

_____

The equations for density and volume of a sphere _____

_____

Hypothesis

I think the smallest/largest ball bearing will fall most quickly through the oil.

Procedure

1. Fill a plastic bottle with vegetable oil.

2. Carefully drop a small, 4 mm ball bearing on the surface and time how long it takes to reach the bottom.

3. Repeat a total of three times.

4. Repeat steps 2 and 3 with 6 mm, 8 mm and 10 mm ball bearings.

28. Ball bearings through oil

Variables

Controlled variable	Controlled variable	Controlled variable
The type of liquid in the tube	The temperature of the liquid	The diameter of the tube
Controlled variable	**Independent variable**	**Controlled variable**
The manner in which the ball bearing is dropped in	The diameter of the ball bearing	
Controlled variable	**Controlled variable**	**Controlled variable**
_____ _____	_____ _____	_____ _____

Dependent Variable – Average speed of the ball bearing.

Results

Calculation: Average speed = Average time (s) ÷ distance ball falls (cm)

Diameter of ball bearing (mm)	Time for ball bearing to reach the bottom (s)				Average speed (cm/s)
	1	2	3	Average	
0					
10					
20					
30					
40					
50					

Graph

Ball bearings through oil

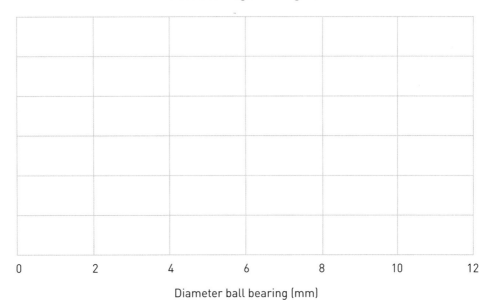

Average speed (cm/s)

Diameter ball bearing (mm)

Discussion

_____

_____

_____

_____

Conclusion

_____

_____

Further Experiment

How do these speeds compare with different temperatures?

Testing the speeds through a variety of liquids.

29.
Electromagnetic strength

Do the number of coils affect the magnetic strength?

TIME REQUIRED – 3 hours

DIFFICULTY – 3 out of 5

DANGERS – Don't connect the wire to the battery for too long as it will heat up and may cause burns or even a fire! Take care with the sharp knife.

Summary

Magnets and magnetism are a hugely important part of modern society. Magnetism is used in motors, speakers and making electricity.

Electromagnets can be made by coiling insulated wire around an iron centre and passing electricity through the wire. Electromagnets can be found in cars, boats, planes, factories and even washing machines!

Materials and Equipment
- 120 cm of insulated copper wire
- Sharp knife
- 6 volt lantern battery
- Large nail
- Pack of pins or paper clips

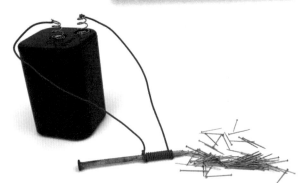

Question

How does changing the number of coils affect the strength of an electromagnet?

Research

What is magnetism? _____

What is an electromagnet? _____

Where are solenoids used? _____

Hypothesis

The more coils of wire, the stronger/weaker the electromagnet will be.

Procedure

1. Carefully strip 2 cm off the plastic insulation from both ends of the wire using the knife.
2. Attach one end of the wire to the positive (+) terminal of the battery.
3. Make 10 tight coils around the nail.
4. Touch the other end of the wire to the negative terminal (don't hold it there for too long because the wire will heat up!).
5. See how many pins can be picked up using the end of the nail.
6. Repeat for 20, 30 and 40 coils.

29. Electromagnetic strength

Variables

Controlled variable	Controlled variable	Controlled variable
The length of the wire	The voltage of the battery	The size of the electrical current
Controlled variable	**Independent variable**	**Controlled variable**
The nail used	The number of coils around the nail	The types of pins being picked up
Controlled variable	**Controlled variable**	**Controlled variable**
_____ _____	_____ _____	_____ _____

Dependent Variable – The strength of the electromagnet.

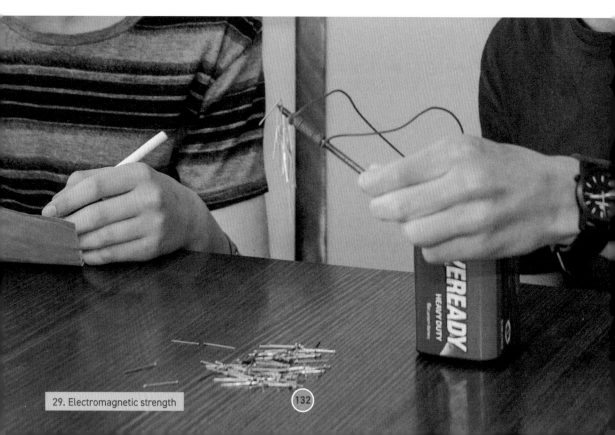

Results

Number of coils	Number of pins picked up
10	
20	
30	
40	

Graph

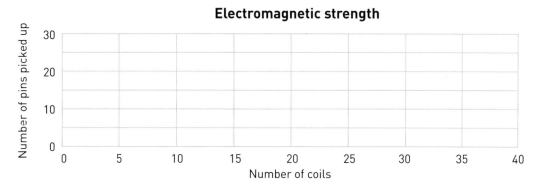

Discussion

_____

_____

_____

_____

Conclusion

_____

_____

Further Experiment

Does the voltage of the battery affect the strength of the electromagnet?

Does the size of the nail matter?

30. Fidget spinner

Find the best angle to hold a spinner

TIME REQUIRED – 3 hours

DIFFICULTY – 3 out of 5

DANGERS – No obvious dangers

Materials and Equipment

- Fidget spinner
- Protractor
- Stop watch

Summary

Fidget spinners were the must have gadget in 2017; almost every kid at school had one! The mesmerising effect of holding a spinning thingy between your thumb and finger seemed to calm and soothe the most difficult student. Why not do an experiment involving these wonderful toys?

Question

What is the best angle to hold a fidget spinner at so that it spins for the longest time unassisted?

Research

What is a fidget spinner? _____

What is momentum? _____

Explore Newton's First Law _____

_____

Hypothesis

The optimum angle to hold a fidget spinner at so that it spins for the longest time unassisted is _____ degrees.

Procedure

1. Hold a fidget spinner flat (angle of 0 degrees) between thumb and index finger.
2. Spin it to the fastest speed you can get it using your other hand.
3. Use the stopwatch to determine how long it keeps spinning for. Repeat.

4. Repeat steps 1–3 by holding it at 15 degrees, 30, degrees, 45 degrees, 60 degrees, 75 degrees and 90 degrees (sideways to ground).

Variables

Controlled variable	Controlled variable	Controlled variable
Controlled variable	**Independent variable**	Controlled variable
Controlled variable	Controlled variable	Controlled variable

Dependent Variable – _____

Results

Angle (degrees)	Time to stop spinning (s)		
	1	2	Average

Graph

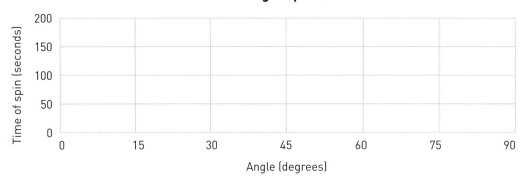

Discussion

_____

_____

Conclusion

_____

_____

Further Experiment

What's the best liquid to lubricate my fidget spinner with?

31. Water bottle rockets

TIME REQUIRED – 4 hours
DIFFICULTY – 4 out of 5
DANGERS – Careful with sharp tools and hot glue gun

Ideal amount of water

Materials and Equipment

- 1.25-litre plastic bottle
- Plastic ice-cream container lids
- Hot glue gun
- Cork with ball inflation needle inserted
- Bicycle pump
- Protractor
- Workshop tools

Summary

Water bottle rockets are fun, but they need a bit of space! In this experiment we explore the optimum amount of water to use in the bottle for the best flight.

Question

What is the best amount of water to use in a water bottle rocket for maximum range?

Research

Newton's Third Law _____

_____

Rockets in general _____

_____

Hypothesis

I think the bottle should be _____ % full of water for the maximum range.

Procedure

1. Make a water bottle rocket like the one shown in the photograph.
2. Fill it with 15% water, angle it at 45 degrees to the ground and launch it.
3. Measure the range. Repeat to get an average.
4. Now repeat with different percentages of water as shown in the results table.

Variables

Controlled variable	Controlled variable	Controlled variable
_____ _____	_____ _____	_____ _____
Controlled variable	**Independent variable**	**Controlled variable**
_____ _____	_____ _____	_____ _____
Controlled variable	**Controlled variable**	**Controlled variable**
_____ _____	_____ _____	_____ _____

Dependent Variable – _____

Results

Percentage of water	Horizontal range (metres)		
	1	2	Average
15			
30			
45			
60			

Graph

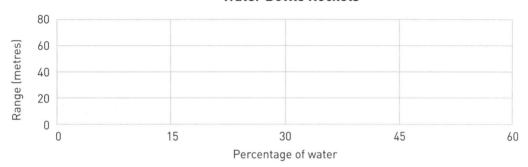

Discussion

_____

_____

_____

_____

Conclusion

_____

_____

Further Experiment

What is the best angle of launch for a water bottle rocket?

Can angled fins make the rocket spin and travel further?

32. Boiling time

Do lids on saucepans help water boil faster?

TIME REQUIRED – 3 hours

DIFFICULTY – 3 out of 5

DANGERS – Be very careful with boiling hot water, make sure an adult is supervising

Materials and Equipment

- A range of different sized saucepans with their lids
- Water
- Stopwatch
- A stove

Summary

When you cook some types of pasta, you sometimes need to get the water boiling in the saucepan first. In this experiment you will test to find out whether it makes a difference to the boiling time if the lid is used.

Question

Does a lid on the saucepan make much of a difference to how quickly the water boils?

Research

Changes of state – liquid to a gas _____

_____

_____

Heat transfer by convection_____

_____

Hypothesis

I think putting a lid on a saucepan will make a little/big difference to how quickly water boils in a saucepan.

Procedure

_____

_____

_____

_____

Variables

Controlled variable	Controlled variable	Controlled variable
Controlled variable	**Independent variable**	**Controlled variable**
Controlled variable	**Controlled variable**	**Controlled variable**

Dependent Variable – _____

Results

Diameter of saucepan (cm)	Height of saucepan (cm)	Time for water to boil without lid (seconds)	Time for water to boil with lid (seconds)

Graph

A column graph will be best to display these results. Put the two results for each saucepan (with and without the lid) side by side.

Discussion

Boiling Times

Boiling time (seconds)

480
420
360
300
240
180
120
60
0

With and without Lid

Conclusion

33. Insulation matters

Find the best insulation to keep a house warm.

TIME REQUIRED – 3 hours

DIFFICULTY – 3 out of 5

DANGERS – No obvious dangers

Materials and Equipment

- 1 small tin can
- 1 large tin can
- A range of insulation materials such as shredded paper, cotton wool, t-shirt material and sand
- Kettle
- Thermometer

Summary

To keep a house warm in winter and cool in summer, there should be insulation in the walls and roof spaces. Do this experiment to find the effectiveness of some different materials.

Question

What common material makes the best insulator?

Research

What is heat?_____

Explain heat transfer by conduction, convection and radiation _____

_____

_____

How do insulation materials reduce heat transfer?_____

_____

Hypothesis

I think the best insulation material will be _____

Procedure

1. Place the small can inside the large can.
2. Stuff shredded paper into the gap between the two cans.

3. Fill the inner can with hot water and record the temperature every 5 minutes for 30 minutes.

4. _____

5. _____

6. _____

Variables

Controlled variable	Controlled variable	Controlled variable

Controlled variable	Independent variable	Controlled variable
_____	Controlled variable	Controlled variable

Dependent Variable – _____

Results

Insulation material	Temperature (degrees Celsius) at each time (minutes)						
	0	5	10	15	20	25	30

Graph

This will be a line graph. You can plot and draw a line for each of the insulation materials.

Record the colour of each of the lines in the key.

Insulation matters

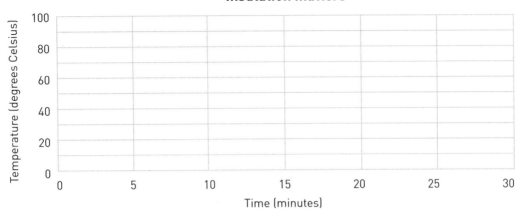

Key for graph

Insulation	Colour of line

Discussion

Insulation	Colour of line

Conclusion

Further Experiment

It would be interesting to test some real insulating materials found at the hardware store.
(Just make sure you take the necessary safety and breathing precautions.)

34. Flying throw tubes
Fun with a plastic bottle

TIME REQUIRED – 3 hours

DIFFICULTY – 3 out of 5

DANGERS – Take care using the scissors

Materials and Equipment
- Scissors
- A variety of sizes of plastic bottles
- Electrical tape
- Measuring tape

Summary

Do you like North American football? If you do, you'll love this experiment! These flying throw tubes are thrown just like a gridiron player passes the ball – one handed, with a spin.

Question

What are the best dimensions for the longest flying throw tube flight?

Research

Bernoulli's effect _____

_____

Stability of spinning objects _____

_____

Hypothesis

As the diameter increases, the flight distance will _____

Procedure

1. Use a pair of scissors to cut the smooth section from the middle of a plastic bottle.
2. Tape one end about five times to get a fairly thick layer.
3. Throw the tube with a spinning motion and the taped end pointing forward. Repeat eight times to find the average distance travelled.
4. Using different bottles make a number of different diameter flyers.

Variables

Controlled variable	Controlled variable	Controlled variable
Controlled variable	**Independent variable**	Controlled variable
Controlled variable	Controlled variable	Controlled variable

Dependent Variable – _____

Results

Diameter (cm)	Distance travelled (m)								
	1	2	3	4	5	6	7	8	Average

Graph

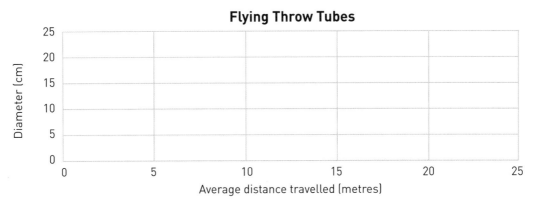

Discussion

_____

_____

_____

Conclusion

_____

_____

Further Experiment

How does changing the length of the tube affect flight range?

Does having more or less tape on the front produce the furthest flight?

35. Fluffy slime!
Finding the best recipe

Summary

Along with fidget spinners, fluffy slime was the must have fad of 2017. Follow this recipe and then do some experimentation to get the best and fluffiest slime ever!

Question

What's the optimum amount of borax solution to add to the fluffy slime to get the best consistency?

Research

What does borax do to PVA glue?

Procedure

1. Add 1 teaspoon of borax to 1 cup of warm water and mix well.

2. Mix the following gently together in a large bowl: ½ cup of white PVA glue, ½ cup of shaving cream, and ½ cup of the foam from foaming hand wash.

3. Sprinkle a tablespoon of cornflour on the mixture and stir in gently with the spatula.

4. Pump in 4 squirts of hand lotion, and gently mix in.

5. Now it's time to experiment! Slowly add the borax solution teaspoon by teaspoon while mixing with the spoon.

6. As the mixture starts getting rubbery, remove it from the bowl and keep mixing with your hands. Keep adding the borax solution teaspoon by teaspoon until you are happy with your slime.

TIME REQUIRED - 2 hours

DIFFICULTY - 3 out of 5

DANGERS - Take care not to get any of the ingredients in your eyes. Keep all ingredients away from young children and do not eat them (the ingredients that is!)

Materials and Equipment

- 1 teaspoon borax
- 1 cup warm water
- ½ cup white PVA glue
- Can of shaving cream
- Bottle of foaming hand wash
- Box of cornflour (cornstarch)
- Bottle of hand lotion
- Food colouring
- A spoon
- A spatula
- A bowl for mixing

Results

Number of teaspoons of borax solution added	Observations, consistency of the slime
0	
1	
2	
3	
4	
5	
6	
7	
8	

Discussion

_____

_____

_____

_____

Conclusion

_____

_____

Further Experiment

What happens to the consistency of the slime when you alter the amounts of some of the other ingredients?

35. Fluffy slime!

36. Rot your teeth

What liquids eat teeth the quickest?

TIME REQUIRED – 2 hours (plus a few days for the liquids to act)

DIFFICULTY – 2 out of 5

DANGERS – No obvious dangers

Summary

Everyone knows that you should brush your teeth before going to bed! But how long does it take for some common foods to start breaking down your teeth? Try an experiment to find out!

Materials and Equipment
- Eggshells
- Drawing chalk (maybe even real teeth if you can get them from the tooth fairy!)
- Vinegar
- Cola
- Lemon juice
- Orange juice
- Lime juice

Question

What effect do different liquids have on teeth?

Research

What are teeth composed of? _____

What are acids?_____

_____

Hypothesis

I think acids will _____

Procedure

Use the list of materials and equipment to plan and write the procedure yourself. (Use the provided results table to give you some hints as well!)

_____

_____

Variables

Controlled variable	Controlled variable	Controlled variable
_____ _____	_____ _____	_____ _____
Controlled variable	**Independent variable**	**Controlled variable**
_____ _____	_____ _____	_____ _____
Controlled variable	Controlled variable	Controlled variable
_____ _____	_____ _____	_____ _____

Dependent Variable – _____

Results

Liquid	Effect each liquid has on the substances below:		
	Eggshell	Chalk	Teeth
Vinegar			
Cola			
Lemon juice			

Discussion

_____

_____

_____

_____

Conclusion

_____

_____

37. The brown apple

The best way to preserve sliced fresh apples

TIME REQUIRED – 2 hours (plus overnight to let experiment proceed)

DIFFICULTY – 2 out of 5

DANGERS – Take care with sharp knife

Materials and Equipment

- Some apples
- A knife
- Chopping board
- 5 bowls
- Water
- Vinegar
- Cola
- Lemon juice
- Orange juice
- Lime juice

Summary

When you peel an apple, it seems to go brown quite quickly. Is there a way to stop this from happening?

Question

What's the best way to stop sliced apples from going brown?

Research

What causes sliced apples to go brown?

_____

_____

Hypothesis

_____

Procedure

Use the list of materials and equipment to plan and write the procedure yourself.

_____

_____

_____

_____

Variables

Controlled variable	Controlled variable	Controlled variable
_____	_____	_____
Controlled variable	**Independent variable**	Controlled variable
_____	_____	_____
Controlled variable	Controlled variable	Controlled variable
_____	_____	_____

Dependent Variable – _____

Results

Use a ruler and plan and draw your own results table.

Discussion

_____
_____

Conclusion

_____
_____

38. Black or silver can

Which heats up quicker?

TIME REQUIRED – 3 hours

DIFFICULTY – 3 out of 5

DANGERS – Use a breathing mask when using spray paint

Summary

Is it better to buy a black or silver car if you live in a hot area? Try this experiment to find out!

Materials and Equipment

- Two aluminium cans
- Black and silver spray paint
- Thermometer
- Water

Question

What heats up quicker in the sun, a black or a silver can?

Research

Why do different objects appear different colours?

Hypothesis

_____

_____

Procedure

Use the list of materials and equipment to plan and write the procedure yourself.
(Use the provided results table to give you some hints as well!)

_____

_____

_____

_____

_____

_____

_____

Variables

Controlled variable	Controlled variable	Controlled variable
_____ _____	_____ _____	_____ _____
Controlled variable	**Independent variable**	**Controlled variable**
_____ _____	_____ _____	_____ _____
Controlled variable	**Controlled variable**	**Controlled variable**
_____ _____	_____ _____	_____ _____

Dependent Variable – _____

38. Black or silver can

Results

Time (minutes)	Black can temperature (degrees Celsius)	Silver can temperature (degrees Celsius)
0		
5		
10		
15		
20		
25		
30		

Graph

Plot a line for each can below.

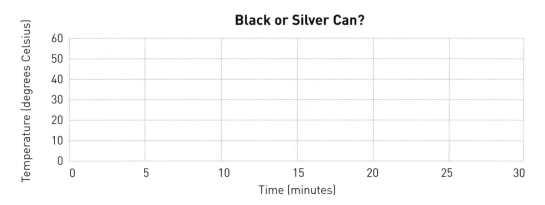

Discussion

_____

_____

_____

Conclusion

_____

_____

39. Pendulum swing

Mass, length and amplitude

TIME REQUIRED – 2 hours

DIFFICULTY – 3 out of 5

DANGERS – No obvious dangers

Materials and Equipment

- 1 metre of string
- A kitchen stool
- 50 grams of Plasticine
- Measuring ruler
- Kitchen scales
- Stopwatch

Summary

Grandfather clocks, metronomes and even children's swings are all based around pendulums. When an object swings back and forth on a string, this is called a pendulum. Some pendulums swing quickly, some slowly. In this experiment you should be able to make some surprising discoveries!

Question

What are the factors which affect the length of time (period) of a pendulum swing?

Research

What is a pendulum? _____

Where are pendulums used? _____

What devices use pendulums? _____

Hypothesis

Factors which affect the period of a pendulum are the mass/length/pull back distance.

• •

Procedure – Effect of changing mass

1. Set up a 30 cm long pendulum with a 10 g mass of Plasticine.
2. Pull it sideways 10 cm and time a total of five full swings. Repeat.
3. Calculate the average of these two times.
4. Now calculate the period (time for one full swing) by dividing your average time by five.
5. Repeat steps 1–4 for a 20 g, 30 g, 40 g and 50 g bob.

Results

Mass of bob (grams)	Time for five full swings (seconds)			Period (seconds)
	1	2	Average	
10				
20				
30				
40				
50				

Discussion

_____

_____

_____

_____

Procedure

Effect of changing the amplitude (how far you pull the bob sideways before releasing it)

1. Set up a 30 cm long pendulum with a 20 g mass of Plasticine.
2. Pull it sideways 5 cm and time a total of five full swings. Repeat.
3. Calculate the average of these two times.
4. Now calculate the period (time for one full swing) by dividing your average time by five.
5. Repeat steps 1–4 for amplitudes of 10 cm, 15 cm and 20 cm.

Results

Initial amplitude (cm)	Time for five full swings (seconds)			Period (seconds)
	1	2	Average	
5				
10				
15				
20				

Discussion

_____

_____

_____

_____

• •

Procedure – Effect of changing the length

1. _____

2. _____

3. _____

4. _____

5. _____

Results

	Time for five full swings (seconds)			Period (seconds)
	1	2	Average	

Graph

Create a line graph with period on the vertical axis and length on the horizontal axis.

Discussion

_____

_____

_____

_____

Conclusion

_____

_____

_____

40. Poisson's law

Water flow and tube diameter

TIME REQUIRED – 3 hours

DIFFICULTY – 4 out of 5

DANGERS – Don't leave buckets of water around as babies and toddlers can drown in them

Materials and Equipment

- A large plastic bucket
- A range of different diameter aquarium hoses (they must all have the same length of approximately 1.5 metres)
- Water
- Stopwatch

Summary

If you've ever had a fishpond or an aquarium then you've probably emptied it at some stage by syphoning the water out using a long hose. Most people think that if you double the diameter of the hose you are using, then you will halve the time it takes to empty the tank. Do this experiment to test whether this is true or not!

Question

What effect does changing the diameter of a siphoning hose have on the time it takes to empty a bucket of water?

Research

Poisson's findings in relation to fluid flow and opening diameter _____

_____

Hypothesis

Doubling the diameter of a siphoning hose will reduce the time of emptying by a factor of

_____

Procedure

1. Fill the bucket with water and then use a tap to fully fill the hose with water.
2. Put a thumb over each end of the hose so that water cannot escape, then put one end in the bottom of the bucket and lay the other end on the ground.

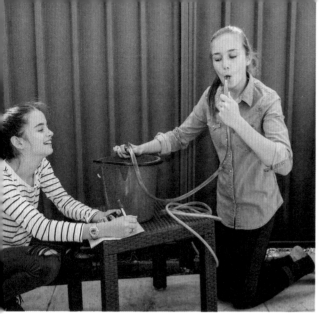 

3. Remove your thumbs and the water should start flowing and emptying the bucket.
4. Repeat for each hose and time how long each of the hoses takes to empty the bucket of water.

Variables

Controlled variable	Controlled variable	Controlled variable
Controlled variable	**Independent variable**	Controlled variable
Controlled variable	Controlled variable	Controlled variable

Dependent Variable – _____

Results

Diameter of hose (mm)	Time to empty bucket (seconds)		
	1	2	Average

Graph

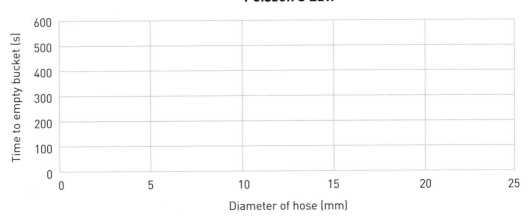

Discussion

_____

_____

_____

_____

Conclusion

_____

_____

41. Rock candy!

How to make the best sugar rock candy

TIME REQUIRED – 3 hours (plus 5 days for sugar crystals to grow)

DIFFICULTY – 3 out of 5

DANGERS – Take care with hot sugar syrup as it can cause burns (adult supervision definitely required!)

Summary

When a super saturated concentration of sugar syrup is made, it readily crystallises to form rock candy. In this experiment you can investigate the different conditions to find the best way to grow rock candy (large sugar crystals).

Materials and Equipment

- 2 cups of water
- 4 ½ cups of sugar
- 6 wooden kebab sticks
- 6 glasses
- 6 pegs
- Saucepan
- Baking paper

Question

What are the best conditions to grow rock candy?

Hypothesis

I think the best place to grow rock candy will be

_____

Research

How do crystals form?_____

_____

Procedure

1. Bring 2 cups of water to a boil in a saucepan.
2. Add ½ cup of sugar at a time, stirring to dissolve.
3. Once no more sugar dissolves, take your saucepan off the heat and allow to cool for 15 minutes.
4. Dip the skewers into the syrup, and then coat the skewers with raw sugar (this will give the rock candy a surface to grow on).
5. Carefully fill each of the glasses with syrup (be careful that no sugar crystals go into the glasses)
6. Lower a sugared kebab stick into the syrup and hold in place with a peg.
7. Loosely place some baking paper over the glass (to keep any nasties out) and put each

glass in a different environmental condition around the house, for example, the fridge, a warm cupboard, a cold cupboard, in the sun etcetera.

Variables

Controlled variable	Controlled variable	Controlled variable
_____ _____	_____ _____	_____ _____
Controlled variable	**Independent variable**	**Controlled variable**
_____ _____	The location where the crystals are put to grow	_____ _____
Controlled variable	**Controlled variable**	**Controlled variable**
_____ _____	_____ _____	_____ _____

Dependent Variable – How well the rock candy grows.

Results

Location	Observations over a few days
Fridge	
Warm cupboard	

Discussion

Conclusion

42. Brine shrimp

TIME REQUIRED – 3 hours (plus a week to hatch the brine shrimp)

DIFFICULTY – 3 out of 5

DANGERS – No obvious dangers

Best amount of salt to hatch your Artemia eggs

Materials and Equipment
- Salt
- Teaspoon
- Water
- Artemia eggs
- Kitchen scales
- Measuring jug
- 6 breakfast bowls

Summary

Sea monkeys (or brine shrimp) used to capture my imagination as a kid. You can buy the eggs (or cysts) online and they arrive in little sachets, ready to hatch!

Question

Is there an ideal salt concentration to hatch Artemia (brine shrimp) eggs?

Research

What are Artemia and how do they reproduce?

_____

_____

Hypothesis

I think the best salt concentration for hatching brine shrimp eggs will be _____ %

Procedure

1. Add ½ teaspoon of salt to a litre of water and dissolve.
2. Fill a bowl with this solution and add some brine shrimp eggs.
3. Repeat steps 1 and 2 with increasing amounts of salt (1, 2, 3 and 4 teaspoons of salt).
4. Set up a control – brine shrimp eggs in a bowl of fresh water (no salt added).
5. Place all the bowls in a warm place and make observations over a few days.

Variables

Controlled variable	Controlled variable	Controlled variable
Controlled variable	**Independent variable**	**Controlled variable**
Controlled variable	Controlled variable	Controlled variable

Dependent Variable – How quickly the brine shrimp hatch.

Results

Number of teaspoons of sugar in 1 litre of water	Observations
0 (control)	
1	
2	
3	
4	
5	

Discussion

_____

_____

_____

_____

_____

Conclusion

_____

_____

_____

42. Brine shrimp

43. Silly putty
How much borax do you add?

TIME REQUIRED – 3 hours

DIFFICULTY – 3 out of 5

DANGERS – Take care not to get any chemicals in your eyes

Summary

What bounces if you throw it on the ground, but stretches if you pull it? Silly putty of course! Every kid loves it – why not do an experiment to find the optimum amount of borax to use?

Question

What is the optimum amount of borax solution for making silly putty?

Materials and Equipment
- Borax powder
- Water
- Medium sized bottle of PVA glue
- Food colouring
- Teaspoon
- Measuring jug
- Mixing bowl

Research

What is silly putty? _____

_____

What does the borax do chemically to PVA? _____

_____

_____

Procedure

1. Dissolve 1 teaspoon of borax in 250 ml of water. This is now the borax solution.

2. Add ¼ cup of white PVA glue to a mixing bowl. Stir in your favourite food colour.

3. Add 1 teaspoon at a time of the borax solution to the mixture and stir in vigorously. Record your observations.

Results

Number of teaspoons of borax solution added	Observations (consistency, viscosity, elasticity, firmness)
0	
1	
2	
3	
4	
5	

Discussion

_____

_____

_____

_____

Conclusion

_____

_____

43. Silly putty

44. Magnetic silly putty

How much iron oxide should you add?

TIME REQUIRED – 3 hours
DIFFICULTY – 3 out of 5
DANGERS – Take care not to get any chemicals in the eyes

Materials and Equipment

- Medium sized bottle of PVA glue
- Borax powder
- 50 grams of iron oxide powder (available online)
- A magnet
- Measuring jug
- Teaspoon
- Mixing bowl

Summary

What's more fun then silly putty? Magnetic silly putty of course! This gloopy monster will swallow magnets just like an amoeba swallows its meals!

Question

What's the best recipe for magnetic silly putty?

Research

What is magnetism? _____

_____

Procedure

Using Experiment 43. Silly Putty as a guide, add various amounts of iron oxide in order to achieve the best magnetic silly putty on planet Earth!

Results

Amount of iron oxide added	Observations (consistency, viscosity, elasticity, firmness, magnetism)

Discussion

_____

_____

_____

_____

Conclusion

_____

_____

45. Flower indicators

Which flowers make the best acid/base indicators?

TIME REQUIRED – 3 hours

DIFFICULTY – 4 out of 5

DANGERS – Take care with hot water as it can cause burns. Adult supervision highly recommended.

Summary

Purple cabbage makes a great acid/base indicator, but so do lots of flowers in the garden!

Materials and Equipment
- A variety of flowers
- Vinegar
- Bicarbonate of soda (baking soda)
- Water
- Saucepan
- Stove
- Sieve
- Drinking glasses
- A spoon

Question

Which flowers make the best acid/base indicators?

Research

What are acids? _____

_____

What are bases? _____

_____

What is an acid/base indicator? _____

_____

Procedure

1. Pick one type of flower from the garden and add the flowers to some boiling water in a saucepan.

2. After a minute of boiling, allow the mixture to cool and use the sieve to filter out bits of flowers, leaving a coloured solution.

3. Test this solution with both vinegar and bicarbonate of soda to see how well it works as an acid/base indicator.

4. Repeat with different flowers as well.

Results

Flower – Name or drawing	Colour with vinegar (acid)	Colour with bicarbonate of soda (base)

Discussion

_____

_____

_____

_____

Conclusion

_____

_____

46. Gummy bear osmosis

Explore osmosis with gummy bears!

TIME REQUIRED – 3 hours (plus a few days for the osmosis to happen)

DIFFICULTY – 2 out of 5

DANGERS – No obvious dangers

Materials and Equipment
- Gummy bears
- 7 drinking glasses
- Salt
- Sugar
- Water

Summary

Osmosis is the diffusion of water from one substance to another. If water diffuses into a gummy bear, then the gummy bear will swell and get bigger.

Question

How does gummy bear osmosis compare in salt and sugar solutions?

Research

Osmosis _____

_____

_____

Hypothesis

_____

Procedure

_____

_____

_____

_____

Variables

Controlled variable	**Controlled variable**	**Controlled variable**
Controlled variable	**Independent variable**	**Controlled variable**
Controlled variable	**Controlled variable**	**Controlled variable**

Dependent Variable – _____

Results (Time to draw your own table!)

Discussion

_____

_____

Conclusion

_____

_____

46. Gummy bear osmosis

47. Plant transpiration

Which plants transpire the most water?

TIME REQUIRED – 3 hours (over a few days)

DIFFICULTY – 3 out of 5

DANGERS – Take care outdoors!

Materials and Equipment

- 10 medium-sized plastic bags
- Bag ties
- Measuring jug

Summary

Plants transpire! This is the process in plants where the sun causes water to be drawn up from the soil and eventually causes it to come out the leaves.

Question

Do all plants transpire the same amount?

Research

Transpiration _____

_____

Hypothesis

_____

Procedure

1. Use a plastic bag to enclose a bunch of leaves on a plant and a bag tie to close off the opening.
2. Repeat step 1 with a number of different types of plants.
3. After a day or so, measure the amount of water collected in each bag.

Variables

Controlled variable	Controlled variable	Controlled variable
_____ _____	_____ _____	_____ _____
Controlled variable	**Independent variable**	Controlled variable
_____ _____	_____ _____	_____ _____
Controlled variable	Controlled variable	Controlled variable
_____ _____	_____ _____	_____ _____

Dependent Variable – _____

Results

Plant	Amount of water collected (ml)

Graph This will be a column graph.

Discussion

_____

_____

_____

Conclusion

_____

_____

48. Bouncy bounce

How does the surface of the ground affect bounce height?

TIME REQUIRED – 2 hours
DIFFICULTY – 3 out of 5
DANGERS – No obvious dangers

Materials and Equipment

- ·
- ·
- ·
- ·

Summary

_____

Question

_____

_____

Research

_____

_____

_____

Hypothesis

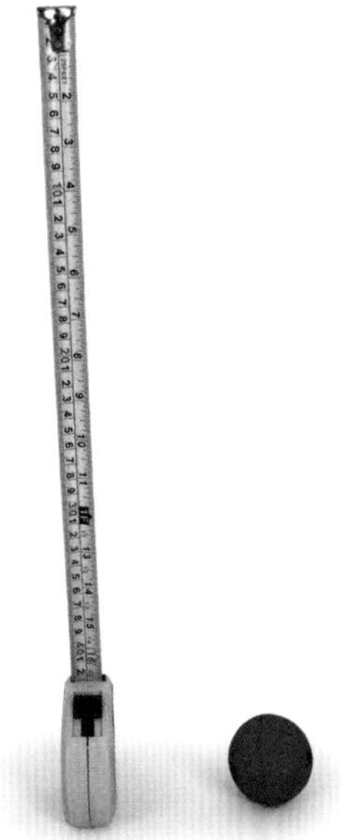

_____

_____

Procedure

_____

_____

_____

_____

Variables

Controlled variable	Controlled variable	Controlled variable
Controlled variable	**Independent variable**	Controlled variable
Controlled variable	Controlled variable	Controlled variable

Dependent Variable – _____

Results

Graph

Discussion

_____

_____

_____

_____

Conclusion

_____

_____

48. Bouncy bounce

49. Yeast growth and temperature

What's the best temperature for yeast growth?

TIME REQUIRED – 2 hours

DIFFICULTY – 4 out of 5

DANGERS – Take care with the hot water

Materials and Equipment

- 6 yeast sachets
- 6 balloons
- 6 bowls
- 6 plastic bottles
- Ice
- Water
- Kettle
- Thermometer

Summary

_____

Question

_____

Research

_____

_____

_____

Hypothesis

_____

_____

Procedure

_____

_____

_____

_____

Variables

Controlled variable	Controlled variable	Controlled variable
Controlled variable	**Independent variable**	Controlled variable
Controlled variable	Controlled variable	Controlled variable

Dependent Variable – _____

Results

Graph

Discussion

_____

_____

_____

Conclusion

_____

_____

49. Yeast growth and temperature

50. Paper helicopter

Factors affecting fall time

TIME REQUIRED – 5 to 50 hours

DIFFICULTY – 4 out of 5

DANGERS – No obvious dangers, but take care if you go somewhere high to release your paper helicopter. Remember to stay safe at all times!

Summary

Paper helicopters are fun to fly so I've put this experiment last as you could literally spend months trying all the different combinations of variables to find the optimum flight time.

Materials and Equipment

- A4 paper
- Pencil
- Ruler
- Pen
- Scissors
- Paperclip
- Measuring tape
- Stopwatch

Question

What are some of the factors which affect the flight time of a paper helicopter?

Research

Investigate weight force _____

_____

What is drag?_____

How does the spinning of a paper helicopter produce lift? _

_____

Hypothesis

The most important variable for a paper glider to remain in the air is _____

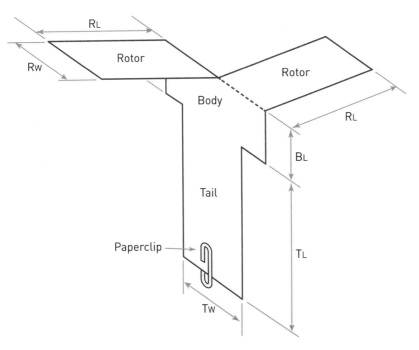

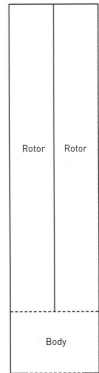

Procedure

1. Make a helicopter glider from paper with the dimensions below:

Label	Dimension	Size (mm)
RL	Rotor Length	100
Rw	Rotor Width	15
BL	Body Length	20
TL	Tail Length	100
Tw	Tail Width	20

2. Release the glider from a balcony and time how long it takes to fly to the ground. Repeat a total of three times.
3. Cut 10 mm off each of the rotor lengths and repeat step 2.
4. Continue shortening the rotor lengths and finding new times of flight.

Results

Rotor Length (mm)	Time of flight (seconds)			
	1	2	3	Average
100				
90				
80				
70				
60				
50				

Discussion

Conclusion

Further Experiment

Now try experimenting with the other variables such as rotor width and tail length.

• •

Final Words

Well I sure hope that you've enjoyed this book! How many experiments did you do? It would be awesome if you shared some photos of you doing the experiments on the book's Facebook page. If you ever get bored and want to watch some fun and entertaining science videos, you should check out my YouTube channel '**Make Science Fun**'.

Anyway, I hope to see you somewhere around the world one day so make sure you say 'Hello!' and don't forget to 'Make Science Fun'!

Jacob